A Practical Guide for Genetic Management of Fragmented Animal and Plant Populations

A Practical Guide for Genetic Management of Fragmented Animal and Plant Populations

Richard Frankham
Macquarie University and Australian Museum, Sydney, NSW, Australia

Jonathan D. Ballou
Species Conservation Toolkit Initiative, Smithsonian Conservation Biology Institute, Washington, DC, USA

Katherine Ralls
Smithsonian Conservation Biology Institute, Washington, DC, USA

Mark D. B. Eldridge
Australian Museum, Sydney, NSW, Australia

Michele R. Dudash & Charles B. Fenster
South Dakota State University, Brookings, SD, USA

Robert C. Lacy
Species Conservation Toolkit Initiative, Chicago Zoological Society, Brookfield, IL, USA

Paul Sunnucks
Monash University, Clayton, Vic, Australia

Line drawings by Karina McInnes, Melbourne, Vic, Australia

OXFORD
UNIVERSITY PRESS

OXFORD

UNIVERSITY PRESS

Great Clarendon Street, Oxford, OX2 6DP,
United Kingdom

Oxford University Press is a department of the University of Oxford.
It furthers the University's objective of excellence in research, scholarship,
and education by publishing worldwide. Oxford is a registered trade mark of
Oxford University Press in the UK and in certain other countries

First Edition published in 2019

Impression: 1

Published in the United States of America by Oxford University Press
198 Madison Avenue, New York, NY 10016, United States of America

British Library Cataloguing in Publication Data
Data available

Library of Congress Control Number: 2019948778

ISBN 978–0–19–878341–1 (hbk.)
ISBN 978–0–19–878342–8 (pbk.)

DOI: 10.1093/oso/9780198783411.001.0001

Printed and bound by
CPI Group (UK) Ltd, Croydon, CR0 4YY

Contents

Contents

Preface

Importance of the topic

One of the greatest unmet issues in conservation biology is the genetic management of fragmented animal and plant populations. Most species across the planet now have fragmented distributions, with many small isolated populations potentially suffering from inbreeding, loss of genetic diversity, and elevated extinction risk. Fortunately, these effects can usually be remedied by initiating gene flow (crossing populations within species), but this is rarely conducted. Disturbingly, evidence of genetic differentiation among populations often is taken to mean that the populations should be kept isolated, thereby dooming many to eventual extinction. We are particularly concerned with populations fragmented by human actions in the last 500 years.

A **paradigm shift** is urgently needed whereby the existence of genetic differentiation among populations acts as a trigger to ask if any populations are suffering from genetic problems (Ralls et al. 2018). If so, as conservation managers we should ask if there are populations to which they can be crossed to reduce inbreeding and increase genetic diversity, and whether this gene flow would be more beneficial than harmful.

Whether to maintain genetic isolation is a major issue for managers of wild animal and plant populations. It is critical that managers, and those who advise them, become aware of the issues relating to restoration of genetic diversity and reversal of genetic deterioration associated with small population size.

Why did we write *A Practical Guide for Genetic Management of Fragmented Animal and Plant Populations*?

We have previously written an advanced textbook (Frankham et al. 2017 *Genetic Management of Fragmented Animal and Plant Populations*) detailing the relevant concepts and supporting evidence. The current *Practical Guide* is shorter, simpler, and more practical than the previous textbook, and is designed for conservation managers and others seeking a briefer and more applied treatment of the topics.

This book provides a concise background to the genetic issues in fragmented wild animal and plant populations and offers means for making practical decisions on genetic management of population fragments. We focus particularly on the harmful consequences of mating among relatives on offspring fitness (inbreeding depression) that are ubiquitous in sexually reproducing animals and plants. A second major focus is on maintenance of genetic diversity, so that species and populations can adapt evolutionarily to changing environments.

Intended audience

This *Practical Guide* is designed for wildlife conservation managers who might have limited familiarity with conservation genetics (conservation ecologists, land and wildlife managers, etc.) and as a concise introduction to the topic for students of conservation biology and wildlife management. To make the book accessible to a wide audience, we have assumed knowledge only of Mendelian genetics and basic statistics. Readers wishing for a simple concise introduction to conservation genetics are referred to *A Primer of Conservation Genetics* (Frankham et al. 2004) or its translations.

Précis of contents

The first five chapters consist of background material and concepts, and the subsequent four chapters are concerned with genetic management (including objectives, questions, and decision trees). The "Take home messages" in the following section summarize the book's messages. This book deals predominantly with outbreeding diploid species, with brief consideration of self-fertilizing and polyploid species. Bold is used for emphasis where necessary.

Historical perspective

A conservation manager we consulted pointed out they previously had avoided implementing genetic rescues on the advice of conservation geneticists, and so sought a historical perspective. Why are we now advising a greater use of genetic rescue? Much of the science we present is recent, especially our recommendations about using genetic rescues. Prior to 2011, outbreeding depression was not considered predictable, so very conservative advice about crossing populations was given (e.g. Edmands 2007). However, Frankham et al. (2011) developed a method for predicting the risk of outbreeding depression (see also Frankham 2015). Further, meta-analyses showed that if there was a low risk of outbreeding depression, genetic rescues led to large fitness benefits that persisted over generations if inbred populations of natural outbreeders were being rescued (Frankham 2015, 2016, 2018). Finally, updated management recommendations were given by Frankham (2015) to reflect these new insights and evidence.

The extent of the fragmentation problem has become ever more serious due to ongoing habitat destruction and fragmentation (IUCN 2018). In earlier times, many remnant populations may have been sufficiently large and genetically diverse, or connected to other populations, to persist. However, this has become increasingly unlikely as populations have continued to decline and become more isolated.

Minimizing mean kinship is widely used for genetically managing fragmented captive populations based on the classic work of Ballou & Lacy (1995). The extension of minimizing kinship to genetic management of fragmented populations in natural habitats has been made possible by recent advances in molecular genetics and genomics that allow estimation of kinship from DNA markers rather than pedigrees. Further, several new insights and results from Ballou and Lacy were presented in Frankham et al. (2017).

Sub-species and species were once regarded as fixed entities, but they are now known to be constantly evolving (Thompson 2013). Further, we now know that hybridization between populations, sub-species, and even species is a normal occurrence in both animals and plants (Coates et al. 2018). Frankham et al. (2012) addressed the conservation implications of different species concepts.

Understanding of the role of climate change in impacting biodiversity has grown markedly, from little mention prior to 2000, to its recognition as one of the most important factors threatening populations and exacerbating genetic problems (Hoffmann & Sgrò 2011; Weeks et al. 2011; Hughes et al. 2017, 2018). Global environmental change makes the need for genetic management of fragmented populations even more pressing (Frankham et al. 2017).

All these insights were reviewed and presented along with genetic management recommendations for fragmented populations in Frankham et al. (2017). The current book presents updated management recommendations that reflect these scientific advances and altered realities faced by conservation managers.

Format

To make the book easy to read and skim, we use a textbook format similar to that of Frankham et al. (2017) *Genetic Management of Fragmented Animal and Plant Populations* and the Frankham, Ballou, & Briscoe textbooks *Introduction to Conservation Genetics* (2002, 2010) and *A Primer of Conservation Genetics* (2004).

Genetic management often involves the use of equations. However, the conservation managers we consulted about this book advised against presenting equations. As a compromise, we have restricted equations to boxes, so they can easily be skipped if preferred. Useful software packages for implementing our recommendations are listed at the end of the chapters.

Referencing

As detailed references were given in our *Genetic Management of Fragmented Animal and Plant Populations*, and due to the length constraints for this guide, we cite only references not given in that book, apart from ones required for copyright purposes. However, the remaining references common to both books are available online http://www.oup.co.uk/companion/FrankhamPG, where they are arranged in chapter groupings for this book. Further, each chapter in this book has about six key sources of **Further reading**, while source references for figures and tables plus new references and Further reading are listed after the Glossary.

New items in the *Practical Guide*

In addition to extensively re-writing the text, re-formulating the headings, making minor changes in structure, incorporating feedback on *Genetic Management of Fragmented Animal Plant Populations*, and drawing on the recommendations of conservation practitioners, we have added the following items not found in the previous book:

- an "Historical perspective" on genetic management of fragmented animal and plant populations (Preface)
- a revised decision tree on genetic management of fragmented populations (Fig. 1.4)
- advice on how to find someone to assist with studying the genetic diversity or the chromosomal constitution of your species (Chapters 2 and 5)
- Box 2.4 on means to obtain an estimate of effective population size (N_e), and Appendix 2 with estimates of N_e/N ratios for a diversity of taxa that can be used to estimate N_e for your population or species
- depiction of genetic differentiation among vertebrate populations that walk, swim, or fly in relation to geographical distances between them (Fig. 4.3)
- explanation of how genetic drift generates random, non-adaptive morphological, behavioral, physiological, and disease resistance variation among populations (Chapter 4)
- potential use of evolutionary rescue to minimize the impact of invasive species (Box 5.3)
- a section on "What not to do in genetic management" (Section II, Table SII.1), added on the recommendation of conservation practitioners
- an unfolding case study "Should isolated, declining populations of Blanding's turtles in central North America be managed as separate populations?" (Box 7.1)
- recommendation that expert opinion should be elicited using recent guidelines that minimize biases in human perception (Chapter 7)
- corrected prediction of benefits of gene flow from lethal equivalents and reduction in inbreeding (Example 7.3)
- prediction of fitness increase due to gene flow when the level of inbreeding depression is known (Example 7.4)
- a revised section on choice of donor population(s) for genetic rescue when there is limited genetic information (Chapter 8)
- a method to estimate the approximate number of generations required until a further round of gene flow into completely isolated population fragments (Example 8.2)
- a section on "Should we manage to conserve rare alleles?" (Chapter 8)
- major revisions of the decision tree for assessing genetic management options to improve the ability of populations and species to cope with global climate change (Fig. 9.2)

- information on a heat-tolerant allele from wild rice (*Oryza australiensis*) that has the potential to be used to improve the heat tolerance of plant species to cope with global climate change (Chapter 9)
- updated information on CRISPR-Cas9 gene editing (Chapter 9)
- six new illustrations of animals and plants by Karina McInnes and one re-used from *Introduction to Conservation Genetics*, 2nd edition
- more than a hundred new references.

Take home messages

1. Genetic management of fragmented populations is one of the most important issues in conservation biology, but is very rarely satisfactorily addressed.
2. Most species now have fragmented distributions, many with small isolated population fragments experiencing loss of genetic diversity, reduced ability to evolve, increased inbreeding, and elevated extinction risks.
3. Harmful effects of inbreeding on fitness (inbreeding depression) are ubiquitous in naturally outbreeding species, and typically cause very large reductions in fitness. If populations of naturally outbreeding species are inbred or have low genetic diversity, they should be presumed to be experiencing inbreeding depression and managed accordingly without waiting for specific evidence of inbreeding depression.
4. Gene flow into an inbred population from another population typically reduces inbreeding and restores genetic diversity and ability to evolve.
5. Crossing between populations occasionally has harmful effects on fitness (outbreeding depression), but the risk is predictable. Crosses between populations that belong to the same species, have the same karyotype, are adapted to similar environments, and have experienced gene flow within the last 500 years have a low risk of outbreeding depression.
6. For genetic management purposes we recommend that species delineations be based on reproductive isolation, as in the Biological Species Concept. Conversely, the Phylogenetic and General Lineage Species Concepts both tend to oversplit, and thus are unsuitable for use with allopatric populations. Additionally, the appropriateness of existing taxonomy should be evaluated before planning a genetic rescue, because many species delineations lack robust scientific support.
7. **We recommend (re)establishing gene flow for isolated population fragments of naturally outbreeding species that are suffering inbreeding or low genetic diversity, provided the proposed population cross has a low risk of outbreeding depression and the predicted benefits justify the cost.** Such gene flow typically results in large improvements in fitness and ability to evolve that persist over generations.
8. When choosing among genetic management actions (and inactions), it is important to assess the overall risks and benefits of different scenarios. **Doing nothing is a choice that is often harmful to the persistence of populations and species.**
9. We recommend managing gene flow among isolated population fragments by minimizing mean kinship. As far as practical, gene flow should be from one or more ecologically similar populations that have the lowest mean kinships with

the recipient population. If kinship analyses are not feasible, we advocate management of gene flow in ways most likely to maximize genetic diversity, based on conservation genetics principles. For example, larger source populations will usually contain more genetic diversity, and ones isolated for longer will typically be more differentiated.

10. Global climate change is increasing the need for genetic management because small populations with little genetic variation will be hindered from adapting to new conditions. Introducing new genetic diversity will decrease their risk of extinction. Where populations with low genetic diversity are translocated to more suitable environments, combining individuals from more than one population will typically increase the chances of success.

11. Threatened species need integrated management across populations, scientific disciplines, institutions, and political boundaries, as exemplified by the One Plan or other similar approaches.

Acknowledgments

We thank our partners Annette Lindsay (RF), Vanessa Ballou (JDB), Robert L. Brownell Jr. (KR), Anne Baker (RCL), and Andrea Taylor (PS) for their forbearance and support during the preparation of this book.

We are grateful for the support of our home institutions, Macquarie University (RF), the Australian Museum (MDBE and RF), the Smithsonian Institution (JDB and KR), South Dakota State University (MRD and CBF), Chicago Zoological Society (RCL), and Monash University (PS).

The authors thank the following for financial research support for projects relevant to the development of ideas presented in the book during the writing period 2017–2019, as follows:

- M. Dudash: her work on inbreeding depression and its genetic basis and how mating system evolution may be impacted was funded through the US National Science Foundation and US Department of Agriculture
- M. Eldridge: his work has been supported by the Australian Research Council, BioPlatforms Australia, and the NSW Environmental Trust
- C. Fenster: his work on the genetic and ecological consequences of population fragmentation has been supported by the US National Science Foundation and Department of Energy (USA), while that on mating system evolution and its genetic consequences has been backed by grants from Natural Sciences and Engineering Research Council of Canada and the Research Council of Norway
- R. Lacy: his work on population management has been supported by various grants from the US National Science Foundation, the Institute of Museum and Library Services, and the Association of Zoos and Aquariums. Much of his work to apply genetic management to species conservation arose from collaborations within the IUCN (International Union for Conservation of Nature) Species Survival Commission, and its Conservation Planning Specialist Group
- P. Sunnucks: his group's work has been supported by sources including the Australian Research Council and the Holsworth Wildlife Research Endowment.

We thank Karina McInnes for preparing the excellent line drawings. Jon Ballou prepared the new Fig. 4.3 in the book and revised many others.

We thank Brian Cypher, Ellen Cypher, Nola Hancock, and Roy Averill-Murray for their helpful suggestions on ways to make the book more practical.

The staff of Oxford University Press and their outsourced production company turned our material into this book. At Oxford University Press, we thank Ian Sherman for championing our book proposal and for overall supervision of the project, and Bethany Kershaw for her efficient and cheerful work as commissioning editor. We thank Cheryl Brant and her team at SPi Global for the effort they put into the production of the book and the obliging way Cheryl addressed our requests for modifications. We are grateful to Julie Musk for the outstanding job she did in copyediting the book.

Copyright acknowledgments

We are grateful to the following for permission to reproduce the material: Chapter 1 frontispiece reprinted from Environmental Policy Law, Vol 6, Oedekoven, K., The vanishing forest, pp 184–185, Copyright (1980), with permission from IOS Press. The publication is available at IOS Press; Box 1.1 figures from Figs 1 and 3 in Zhang, B., Li, M., Zhang, Z., Goossens, B., Zhu, L., Zhang, S., Hu, J., Bruford, M.W., Wei, F., 2007. Genetic viability and population history of the giant panda, putting an end to the "evolutionary dead end"? Molecular Biology and Evolution 24(8), 1801–1810, with permission of Oxford University Press; Chapter 3 frontispiece plot (b) and Fig. 3.3 republished with permission of John Wiley and Sons Inc. from Figs 3 and 4 from Szulkin, M., Garant, D., McCleery, R.H., Sheldon, B.C., 2007. Inbreeding depression along a life-history continuum in the great tit. Journal of Evolutionary Biology 20(4), 1531–1543, permission conveyed through Copyright Clearance Center, Inc; Chapter 3 frontispiece plot (c) republished with permission of John Wiley and Sons Inc. from Fig. 2 in Nason, J.D., Ellstrand, N.C., 1995. Lifetime estimates of biparental inbreeding depression in the self-incompatible annual plant *Raphanus sativus*. Evolution 49(2), 307–316, permission conveyed through Copyright Clearance Center, Inc.; Chapter 4 frontispiece map from Fig. 2 in James, F., 1995. The status of the red-cockaded woodpecker and the prospect for recovery. In: Red-Cockaded Woodpecker: Recovery, Ecology and Management, eds D.L. Kulhavy, R.G. Hopper, R. Costa. Center for Applied Studies, Stephen F. Austin State University, Nacogdoches, TX.; Fig. 4.2 reprinted with permission from Taylor & Francis Ltd, from Cremer, K.W., 1966. Dissemination of seed from *Eucalyptus regnans*. Australian Forestry 30(1), 33–37, http://www.informaworld.com; Fig. 4.3 from Fig. 1b in Medina, I. Cooke, G.M., Ord, T. J., 2018. Walk, swim or fly? Locomotor mode predicts genetic differentiation in vertebrates. Ecology Letters 21, 638–645, © 2018 John Wiley & Sons Ltd/CNRS; Fig. 4.4 from Fig. 5 in Aguilar, R., Quesada, M., Ashworth, L., Herrerias-Diego, Y., Lobo, J., 2008. Genetic consequences of habitat fragmentation in plant populations: susceptible signals in plant traits and methodological approaches. Molecular Ecology 17, 5177–5188, © 2008 Blackwell Publishing Ltd; Fig. 4.11 from Fig. 1 in Coulon, A., Fitzpatrick, J.W., Bowman, R., Stith, B.M., Makarewich, C.A., Stenzler, L.M., Lovette, I.J., 2008. Congruent population structure inferred from dispersal behaviour and intensive genetic surveys of the threatened Florida scrub-jay (*Aphelocoma cœrulescens*). Molecular Ecology 17(7), 1685–1701, doi: 10.1111/j.1365-294X.2008.03705.x, © 2008 Blackwell Publishing Ltd; Fig. 5.1 from Fig. 1 in Newman, D., Tallmon, D.A., 2001. Experimental evidence for beneficial fitness effects of gene flow in recently isolated populations. Conservation Biology 15(4), 1054–1063, published by Wiley for Society for Conservation Biology; Fig. 5.2 from Fig. 2 in Frankham, R., 2015. Genetic rescue of small inbred populations: meta-analysis reveals large and consistent benefits of gene flow. Molecular Ecology doi: 10.1111/mec.13139, Copyright 1999–2016 John Wiley & Sons, Inc.; Fig. 5.3 from Fig. 3 in Murray, B., Young, A.G., 2001. Widespread chromosome variation in the endangered grassland forb *Rutidosis leptorrhynchoides* F. Muell. (Asteraceae: Gnaphalieae). Annals of Botany 87(1), 83–90, republished with permission of Oxford University Press; Fig. 6.2 from Figs 1 and 3 in Bouzat, J.L., Cheng, H.H., Lewin, H.A., Westemeier, R.L., Brawn, J.D., Paige, K.N., 1998. Genetic evaluation of a demographic bottleneck in the greater prairie chicken. Conservation Biology 12(4), 836–843, published by Wiley for Society for Conservation Biology; Chapter 7 frontispiece from Fig. 2 in Oakley, C.G., Winn, A.A., 2012. Effects of population size and isolation on heterosis, mean fitness, and inbreeding depression in a perennial plant. New Phytologist 196(1), 261–270, © 2012 The Authors. New Phytologist © 2012 New Phytologist Trust; figures in Box 7.2 from Figs 1, 4, and 5 from Coleman, R.A., Weeks, A.R., Hoffmann, A.A., 2013. Balancing genetic uniqueness and genetic variation in determining conservation and translocation strategies: a comprehensive case study of threatened dwarf galaxias, *Galaxiella pusilla* (Mack) (Pisces: Galaxiidae). Molecular Ecology 22(7), 1820–1835, © 2013 Blackwell Publishing Ltd; image in Box 7.3 reprinted from Roelke, M.E., Martenson, J.S., O'Brien, S.J., 1993. The consequences of demographic reduction and genetic depletion in the endangered Florida panther. Current Biology 3(6), 340–350, Copyright 1993, with permission from Elsevier; Example 8.4 from Fig. 1 in Finger, A., Kettle, C.J., Kaiser-Bunbury, C.N., Valentin, T., Doudee, D., Matatiken, D., Ghazoul, J., 2011. Back from the brink: potential for genetic rescue in a critically endangered tree. Molecular Ecology 20(18), 3773–3784, © 2008 Blackwell Publishing Ltd.

List of symbols

The following table is a list of symbols used in the book, along with their meaning and the chapters where they are mainly considered (after Frankham et al. 2017 pp. xix–xxiii). Different symbols for the same variable are sometimes used in diverse publications.

Symbol	Definition	Chapter
A	Allelic diversity, mean number of alleles per genetic marker or gene	2
B	Haploid lethal equivalents	3, 7
f	Relative frequency	7
F	Inbreeding coefficient of individuals, or the mean for a population	2, 3, 5, 7, 8
F_{IS}	Inbreeding within population fragments	4
F_{IT}	Total inbreeding over a number of population fragments	4
F_{ST}	Inbreeding due to population differentiation	4, 8
H	Heterozygosity (see also H_e H_o, H_S, and H_T)	2, 3, 4, 5, 7, 8
H_e	Hardy–Weinberg equilibrium heterozygosity	2, 4, 7
H_o	Observed heterozygosity	2, 4
H_S	Expected heterozygosity averaged across population fragments	4
H_T	Expected heterozygosity across the totality of a group of population fragments treated as a single unit	4
ID	Inbreeding depression, difference in mean between outbred and inbred offspring for a quantitative trait (e.g. fitness)	3
k_{ij}	Kinship (or coancestry) between two individuals: the probability that two randomly chosen alleles at a locus, one from each individual, are identical by descent	4, 8
ln	Natural logarithm, $\log_e$	3, 7, 8
m	Migration rate, proportion of migrant alleles introduced over a single generation, where Nm is the number of migrants	8
$\overline{mk}$	Average kinship of a population	8

Symbol	Definition	Chapter
mk_{AB}	Mean kinships between populations A and B, average of pairwise kinships of all individuals in population A with all in population B	4, 8
mk_i	Mean kinship of individual i, average of its kinships with all individuals in the population	8
n	Sample size	2, 5
N	Census population size (usually potentially reproducing adults)	2, 4, 8, 9
N_e	Effective population size. This determines the genetic impacts of small population size	2, 3, 4, 7, 8, 9
p	Frequency of allele A_1 (or A)	2
q	Frequency of allele A_2 (or a)	2
s	Selection coefficient against a genotype at a locus	2
t	Number of generations	4, 7, 8
W	Mean fitness of a population or group	7
δ	Measure of inbreeding depression: proportionate decrease in trait mean due to inbreeding compared to outbreeding	3
Δ	Change in value, usually from one generation to the next (see also ΔF, ΔGR, and ΔW)	5, 7
ΔF	Differences in inbreeding coefficient between individuals in different generations	5, 7
ΔGR	Genetic rescue effect, measured as the proportionate benefit in fitness of a crossed population compared to the inbred parent	5
ΔW	Change in fitness, e.g. due to inbreeding	7

Introduction

CHAPTER 1

Genetic management of fragmented populations is one of the major, largely unaddressed issues in biodiversity conservation. Many species across the planet have fragmented distributions with small isolated populations that are potentially suffering from inbreeding and loss of genetic diversity (genetic erosion), leading to elevated extinction risk. Fortunately, genetic deterioration can usually be remedied by gene flow between populations, yet this is rarely done, in part because crossing is sometimes harmful (outbreeding depression). However, it is now possible to predict when this problem might occur. We consider means for determining whether genetic problems might be alleviated by crossing, evaluating the risks of outbreeding depression and the potential benefits of gene flow, and for optimizing gene flow. We also consider genetic management under global climate change.

TERMS

Critically endangered, endangered, gene flow, genetic diversity, genetic erosion, genetic rescue, inbreeding, inbreeding depression, outbreeding depression, population, population fragmentation, threatened

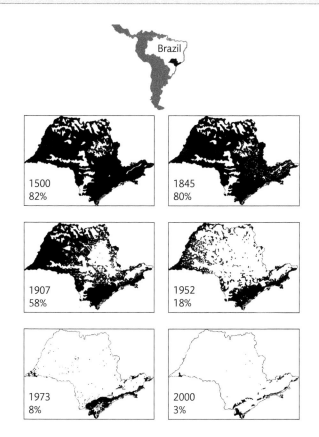

Loss and fragmentation of the Atlantic coastal forest of Brazil (black region on top map) from 1500 to 2000 (from Oedekoven 1980) indicating that the distributions of many species have become increasingly fragmented. The percentages are extent of forest cover.

A Practical Guide for Genetic Management of Fragmented Animal and Plant Populations. R. Frankham, J. D. Ballou, K. Ralls, M. D. B. Eldridge, M. R. Dudash, C. B. Fenster, R. C. Lacy & P. Sunnucks. Oxford University Press (2019). © R. Frankham, J. D. Ballou, K. Ralls, M. D. B. Eldridge, M. R. Dudash, C. B. Fenster, R. C. Lacy & P. Sunnucks 2019. DOI: 10.1093/oso/9780198783411.001.0001

Why do we need genetic management of species?

Inbreeding and loss of genetic diversity are unavoidable in small isolated populations. Inbreeding reduces reproduction and survival and thus increases extinction risk. Loss of genetic diversity reduces the ability of populations and species to evolve, and thereby increases extinction risk, especially under environmental change.

The IUCN (International Union for Conservation of Nature), the premier international conservation agency, recommends conservation of biodiversity at three levels:

- genetic diversity
- species diversity
- ecosystem diversity.

Genetic issues are involved in all three levels. First, genetic diversity is worthy of conservation in its own right because it controls the variety of life—the shape, size, color, biochemistry, behavior, and all other characters that define species and underpin their place in the world. Second, adequate genetic diversity favors species persistence by defending against harmful effects of inbreeding, maintaining fitness, and providing the ability to evolve. Third, through these influences on species diversity, genetic diversity enhances ecosystem viability.

Genetics has received little attention in the management of threatened species in natural habitats, despite genetic diversity being protected under international conventions, such as the United Nations Convention on Biological Diversity, and national laws in many countries. While genetic threats to persistence and recovery were mentioned in 63% of endangered species recovery plans in the USA, 55% in Australia, and 33% in Europe, few of these mentioned active genetic management. Further, inbreeding was commented upon only in ~ 7% of plans and even fewer discussed the harmful effect of inbreeding on fitness (inbreeding depression) (see also Cook & Sgrò 2017), even though examples of the consequences of ignoring genetic issues and poor handling of them are common in wildlife management. In contrast, genetics has played a major role in the management of captive populations since the 1980s.

Efforts have recently been made to incorporate genetic considerations into the management of wild populations, such as a proposal for conservation genetic monitoring in Sweden, and creation of the Conservation Genetics Specialist Group within the IUCN Species Survival Commission.

However, much more needs to be done. In our view, **the most important genetic issue in conservation biology is the genetic management of fragmented animal and plant populations**, and it is largely unaddressed. This book is designed to provide conservation practitioners with the methodology to conserve genetic diversity through the genetic management of fragmented animal and plant populations.

Ubiquity of fragmented populations, some suffering genetic problems

Most species on the planet have fragmented distributions, much of it due to human destruction and modification of natural habitats over the last few hundred years. This includes many threatened animal and plant species on the IUCN Red List. Many isolated populations have reduced genetic diversity relative to historic levels. For example, an estimated 26% of invertebrate, 29% of vertebrate, and 55% of plant species show evidence of inadequate gene flow among populations.

Inbreeding depression and loss of genetic diversity will ultimately contribute to the extirpation of many small populations of sexually reproducing organisms (see later in this chapter, and Chapters 3 and 4). Box 1.1 illustrates the fragmented distribution of the giant panda (*Ailuropoda melanoleuca*) and some of its genetic consequences.

Box 1.1 Fragmented distribution of the giant panda in China and its genetic consequences

(after Frankham et al. 2017, Box 1.1, based on Zhang et al. 2007)

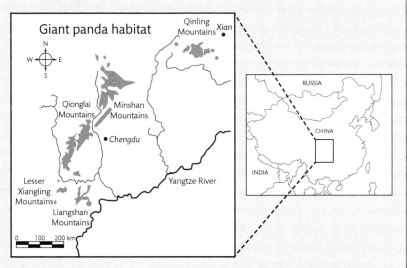

Giant panda (China)

Giant pandas in China have a geographically fragmented distribution (left map). Genetic analyses have shown that the most isolated populations in the Qinling and Liangshan Mountains have lower genetic diversity than other larger, less physically isolated populations, such as those in the Minshan and Qionglai Mountains, based on microsatellite markers (short tandem repeat: Appendix 1).

Population	Heterozygosity	Allelic diversity
Qinling (QIN)	0.486	2.58
Liangshan (LIS)	0.366	2.30

Continued ➢

Population	Heterozygosity	Allelic diversity
Minshan (MIN)	0.559	3.03
Qionglai (QIO)	0.610	3.18
Lesser Xiangling (XL)	0.635	3.29

The populations have diverged genetically due to isolation and now fall into four different genetic clusters, QIN, MIN, QIO, and LIS plus XL as shown in the STRUCTURE plot below (Zhang et al. 2007). Bars represent the genotypes of individual pandas. Individuals with genotypes typical of populations other than their own are likely immigrants, while those with genotypes intermediate between populations indicate gene flow. There is some gene flow between the populations, but very little into Qinling.

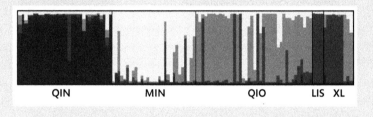

Because the term "population" is used widely in this book, we now define it in a conservation genetics context.

What is a population?

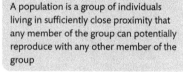

A population is a group of individuals living in sufficiently close proximity that any member of the group can potentially reproduce with any other member of the group

The above definition is based on evolutionary considerations, as appropriate for the issues in this book.

There is a continuum between completely isolated populations and completely connected ones, depending on levels of gene flow and the number of generations of separation (Fig. 1.1).

Fig. 1.1 The continuum of genetic differentiation among populations (after Frankham et al. 2017, Fig. 1.1, based on Waples & Gaggiotti 2006). Each group of circles represents a group of population fragments with varying degrees of connectivity (geographic overlap or gene flow). (a) Connected populations where any individual can potentially mate with any other. (b) Substantial connectivity. (c) Modest connectivity. (d) Complete genetic isolation.

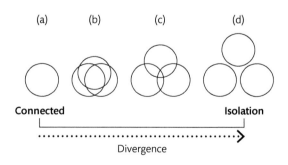

Genetic problems are typically worse in fragmented than continuous populations

Fragmented populations with inadequate gene flow among them suffer from the following genetic problems:

- inbreeding and inbreeding depression
- loss of genetic diversity and reduced ability to adapt evolutionarily
- elevated extinction risk.

Small isolated populations of outbreeding species suffer unavoidable loss of genetic diversity and inbreeding over generations with consequent fitness reductions and compromised ability to evolve. The adverse genetic changes occur at a rate that is inversely proportional to the population size (Chapter 2). When species are fragmented into genetically isolated units, the rate of genetic deterioration accelerates because it is determined by the population size of the fragment, not the whole species. Over generations, these genetic problems accumulate in isolated populations and contribute to population extinctions (Fig. 1.2), often in concert with other threats such as human impacts, and demographic and environmental fluctuations. Further, genetic problems are typically worse when environmental conditions are changing rapidly (Chapter 9).

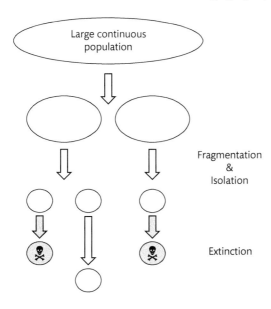

Fig. 1.2 Stages over time (top to bottom) from one large continuous population, through fragmentations into two substantial populations, to three small populations, followed by two extinctions, resulting in one small surviving population (Frankham et al. 2017, Fig. 1.2).

Why conserve isolated populations?

Genetic diversity is distributed among species, populations within species, and individuals within populations. Genetic diversity is lost through random sampling of gametes to produce successive generations in finite populations (Chapter 2), selection

Conservation of geographically separate populations is important for conserving species' genetic diversity, hedging against catastrophes, maintaining suitable habitat, and minimizing species extinction risk

favoring some alleles over others, and extinction of populations. As species are composed of populations, extirpations of populations are steps towards species extinction (Ceballos et al. 2017). Geographically dispersed populations of a species represent a buffer against extinction risk from local catastrophes, help retain genetic diversity, and provide a sound justification for conserving habitat in the face of pressure from human activities.

The adverse effects of population fragmentation can usually be reversed by gene flow

Inbreeding depression and loss of genetic diversity can often be reversed by establishing gene flow between populations, referred to as **genetic rescue**, often with large fitness benefits (Chapter 5). For example, gene flow substantially improved the fitness of small inbred populations of organisms as diverse as African lions (*Panthera leo*), Glanville fritillary butterflies (*Melitaea cinxia*), and the critically endangered jellyfish tree (*Medusagyne oppositifolia*) (Fig. 1.3). Gene flow sometimes has harmful fitness consequences, but such cases are uncommon, predictable, and populations usually recover in the following generations (Chapter 5).

Fig. 1.3 Genetic rescue due to gene flow into small inbred populations of African lions, Glanville fritillary butterflies on a Russian island, and critically endangered jellyfish trees from the Seychelles. Composite fitness (reproduction plus survival) was measured in the lions and trees and egg hatchability in the butterflies (after Frankham et al. 2017, Fig. 1.3, based on Trinkel et al. 2008; Finger et al. 2011; Mattila et al. 2012).

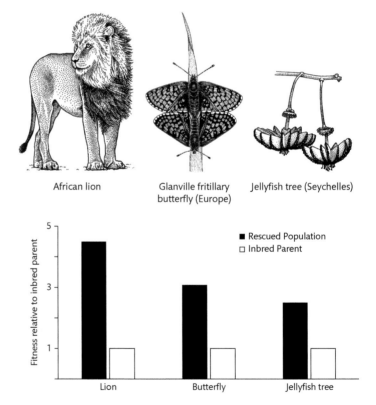

African lion   Glanville fritillary butterfly (Europe)   Jellyfish tree (Seychelles)

Importance of genetically managing fragmented populations

Most species have fragmented distributions, many with isolated population fragments suffering genetic problems, but such genetic problems are usually neglected, and the populations left to go extinct. Although most isolated inbred populations could be saved from extinction by crossing with another population, this is very rarely done. For example, the Isle Royale population of the gray wolf (*Canis lupus*) was highly inbred and suffering severe skeletal problems, but the US National Parks Service in 2014 refused to augment the island population from the mainland and it declined to a single highly inbred pair by 2016, but in 2018 the authorities relented (Matheny 2018).

> Effective genetic management of fragmented populations is one of the major, largely unaddressed issues in conservation biology

Many populations would potentially benefit from gene flow

All species with fragmented distributions that have at least one totally isolated population and at least one other population can potentially benefit genetically from gene flow. The smaller the population sizes of isolated fragments and the more generations they have been isolated, the greater the need to establish gene flow to increase genetic diversity and reduce inbreeding.

Examples of species that have isolated populations and would likely benefit from gene flow include greater one-horned rhinoceroses (*Rhinoceros unicornis*) in India and Nepal; black-footed rock-wallabies (*Petrogale lateralis*), koalas (*Phascolarctos cinereus*), Cunningham's skinks (*Egernia cunninghami*), and matchstick banksias (*Banksia cuneata*) in Australia; giant pandas in China; leopards (*Panthera pardus*) in South Africa; grizzly bears (*Ursus arctos horribilis*), desert topminnow fish (*Poeciliopsis monacha*), scarlet gilia (*Ipomopsis aggregata*), and swamp pink (*Helonias bullata*) plants in North America; tuataras (*Sphenodon punctatus*) and several species of birds in New Zealand; and Glanville fritillary butterflies in Russia.

Genetic problems due to population fragmentation are estimated to be affecting ~ 1.4 million isolated populations of threatened species. Considering all species, this number climbs to ~ 150 million isolated populations likely manifesting genetic problems.

Few genetic rescues have been attempted

We are aware of only ~ 30 cases where gene flow from non-locally derived populations may have been implemented for conservation purposes worldwide (Table 1.1). These cases represent a miniscule proportion of the small fragmented populations that could benefit from management of gene flow.

We are not aware of any other unaddressed scientific issue in conservation biology that could have such beneficial practical effects on species conservation in the wild as would wise application of genetic rescues and, furthermore, at relatively modest cost.

Table 1.1 Threatened species and populations where gene flow has been augmented for known or potential conservation purposes to alleviate genetic problems (known or suspected) (updated from Frankham et al. 2017, Table 1.1).

Common name	Taxon
Mammals	
African elephant	*Loxodonta africana*
African lion	*Panthera leo*
African wild dog	*Lycaon pictus*
Black rhinoceros	*Diceros bicornis*
Bighorn sheep	*Ovis canadensis*
Black-footed rock-wallaby	*Petrogale lateralis*
Brush-tailed rock-wallaby	*Petrogale penicillata*
Burrowing bettong[a]	*Bettongia lesueur*
Columbia Basin pygmy rabbit	*Brachylagus idahoensis*
Florida panther	*Puma concolor coryi*
Golden lion tamarin	*Leontopithecus rosalia*
Mexican wolf	*Canis lupus baileyi*
Mountain pygmy possum	*Burramys parvus*
Northern quoll	*Dasyurus hallucatus*
Nubian giraffe[b]	*Giraffa camelopardalis camelopardalis*
Proserpine rock-wallaby	*Petrogale persephone*
Western barred bandicoot[a]	*Perameles bougainville*
Birds	
Greater prairie chicken in Illinois	*Tympanuchus cupido pinnatus*
Helmeted honeyeater[c]	*Lichenostomus melanops cassidix*
Red-cockaded woodpecker	*Picoides borealis*
Reptiles	
Bojer's skink[d]	*Gongylomorphus bojerii*
European adder	*Vipera berus*
Fish	
Macquarie perch[e]	*Macquaria australasica*
Plants	
Beach clustervine	*Jacquemontia reclinata*
Brown's banksia	*Banksia brownii*
Button wrinklewort daisy	*Rutidosis leptorrhynchoides*
Florida ziziphus	*Ziziphus celata*

Common name	Taxon
Lakeside daisy population in Illinois	*Hymenoxys acaulis* var. *glabra*
Mauna Kea silversword	*Argyroxiphium sandwicense* ssp. *sandwicense*
Marsh grass of Parnassus	*Parnassia palustris*
Round-leafed honeysuckle	*Lambertia orbifolia*
Spiny daisy	*Acanthocladium dockeri*
Torrey pine[f]	*Pinus torreyana*
Twinflower	*Linnaea borealis*

[a] White et al. (2018);
[b] Fessessy, J., pers. comm.;
[c] Harrisson et al. (2016);
[d] du Pleissis et al. (2018);
[e] Pavlova et al. (2017);
[f] Hamilton et al. (2017).

Why have so few genetic rescues been attempted?

The limited use of genetic rescue is due to:

- fears of outbreeding depression
- limited quantitative information on the fitness effect of gene flow
- overly stringent guidelines for attempting genetic rescue
- concerns over loss of genetic purity, identity, and adaptation to local conditions
- limited access to genetic expertise and funding.

Outbreeding depression is a reduction in fitness (reduced ability to mate/pollinate, fertilize, produce offspring, and survive) following crossing of populations. Fear of outbreeding depression is frequently a major impediment to genetic rescue attempts. Thus, it is critical to assess the risk of outbreeding depression in crosses that might otherwise help fragmented populations. Outbreeding depression occurs in only some crosses, and there are now practical means to predict this risk (Chapters 5 and 7).

Recent meta-analyses showed that gene flow into inbred populations produced large consistent fitness benefits in F_1, F_2, and F_3 and later generations, provided that crosses with an elevated risk of outbreeding depression were excluded (Chapter 5).

Until recently, overly stringent guidelines for genetic rescue likely discouraged managers of wild populations from attempting it. We provide appropriate, updated guidelines based on recent scientific advances (Chapter 7). Loss of local adaptation is also usually a minor and manageable issue that can be addressed by gene flow from ecologically similar populations (Chapters 5 and 7).

Limited access to genetic expertise and funding were identified as major impediments to genetic management of threatened species in a survey of conservation practitioners in New Zealand (Taylor et al. 2017). This book is designed to alleviate the expertise problem.

How should we genetically manage fragmented populations?

In this book, we provide practical guidelines for genetically managing fragmented populations. These involve (a) determining whether populations are suffering genetic problems, and if so (b) whether there is another population to which it could be crossed, (c) whether gene flow would be beneficial or harmful, and (d) how much gene flow is appropriate and from where

A flowchart of the questions we need to answer to genetically manage fragmented animal and plant populations is given in Fig. 1.4. The background to these questions is addressed in Chapters 2–5. Chapter 6 deals with appropriate delineation of species for conservation purposes. Chapters 7–9 deal with developing management decisions. In Chapter 7 we ask whether any isolated fragments are suffering inbreeding depression, loss of genetic diversity, and elevated extinction risks (genetic erosion), and whether they would benefit from augmented gene flow. Managing gene flow by moving individuals or gametes (Chapters 8) involves answering the questions:

- From where?
- How many?
- What population genetic statistics should we manage?
- Which individuals?
- How often?
- When should we cease?

Chapter 9 considers how global climate change increases the need for genetic management, and the book concludes with the "Final messages for managers of wild animal and plant populations". The overall content is summarized in the "Take home messages" that appeared earlier. We emphasize that genetic management issues must be integrated with those of other disciplines into a single comprehensive plan that includes all stakeholders and relevant political jurisdictions, as in the One Plan Approach developed by the IUCN's Conservation Planning Specialist Group.

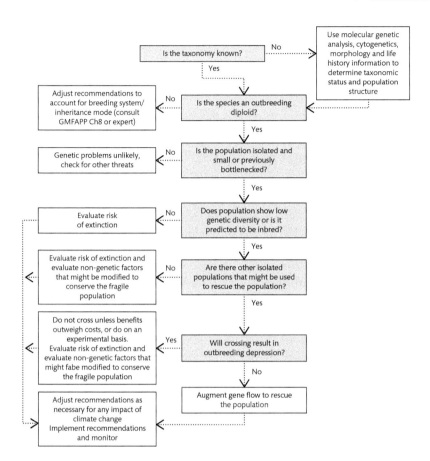

Fig. 1.4 Decision tree with the main questions that need to be asked when genetically managing fragmented populations (revised from Frankham et al. 2017, Fig. 1.4).

In the next section, we present background material on genetic problems in small isolated populations. The first of these chapters provides a brief introduction to the evolutionary genetics of small populations, as a prelude to dealing in more detail with genetic problems in small populations.

Summary

1. Instituting appropriate genetic management of fragmented populations is one of the major issues in conservation biology that is largely unaddressed.
2. Many species have isolated populations that are suffering loss of genetic diversity, inbreeding, and elevated extinction risks: many of these would benefit from gene flow.
3. There are very few cases where genetic rescue has been attempted.
4. Fear of outbreeding depression, limited information on benefits of gene flow, the desire to maintain genetic purity, and overly stringent guidelines have impeded appropriate genetic management of fragmented populations. None of these represents insurmountable problems, as we show in later chapters.

5. We present background material for understanding inbreeding and outbreeding depression, and provide practical guidelines for identifying genetic problems in fragmented populations. Further, we consider how to assess whether genetic problems might be alleviated by instituting gene flow, evaluate the risks of outbreeding depression, and implement appropriate gene flow regimes.

6. Management of fragmented populations under global climate change will require even more attention to genetic issues, as doing nothing will be less and less likely to be a "safe" option.

7. Genetic issues need to be integrated with those from other disciplines.

FURTHER READING

Frankham et al. (2017) *Genetic Management of Fragmented Animal and Plant Populations*: Chapter 1 provides a more detailed treatment of topics in this chapter.

Pierson et al. (2016) Provides an evaluation of how often genetic factors were considered in endangered species recovery plans from Australia, Europe, and the USA.

Ralls et al. (2018) Makes a call for a paradigm shift in the genetic management of fragmented populations.

SECTION I

Genetic problems in small isolated populations and their remedies

We begin this section with a brief introduction to evolutionary genetics of small populations (Chapter 2), especially for readers who are not familiar with entry-level conservation genetics or evolutionary genetics. In this chapter, we consider the roles of mutation, gene flow, selection, and chance in evolution, and describe the magnitude of genetic diversity in populations that results from these factors. Subsequently, we define genetically effective population size and inbreeding, and how they are measured.

Chapter 3 addresses genetic problems that occur within species as a consequence of small populations, namely:

- inbreeding depression
- loss of genetic diversity, resulting in reduced ability to evolve.

Fragmentation of populations is a major problem when it leads to inadequate gene flow, because problems due to inbreeding and loss of genetic diversity are worsened, and extinction risk elevated (Chapter 4). If population fragments are totally isolated, then the genetic fate of each fragment is determined by its effective population size, rather than that of the whole species.

The adverse impacts of inbreeding and loss of genetic diversity on fitness and ability to evolve are usually reversible by gene flow from another population (Chapter 5). Despite its expected benefits, genetic rescue has rarely been attempted for conservation purposes, in part through lack of relevant information. Fortunately, such information is now available. In Chapter 5 we also identify chromosomal differences and adaptation to different environments as factors that predict the risk of harmful effects of crossing, allowing us to remove any impediment to the use of genetic rescue.

The "**List of Symbols**," located after the "Acknowledgments," details the symbols used in this book and the chapters where they are mainly considered.

A Practical Guide for Genetic Management of Fragmented Animal and Plant Populations. R. Frankham, J. D. Ballou, K. Ralls, M. D. B. Eldridge, M. R. Dudash, C. B. Fenster, R. C. Lacy & P. Sunnucks. Oxford University Press (2019). © R. Frankham, J. D. Ballou, K. Ralls, M. D. B. Eldridge, M. R. Dudash, C. B. Fenster, R. C. Lacy & P. Sunnucks 2019. DOI: 10.1093/oso/9780198783411.001.0001

Evolutionary genetics of small populations

CHAPTER 2

Genetic management of fragmented populations involves the application of evolutionary genetic theory and knowledge to alleviate problems due to inbreeding and loss of genetic diversity in small populations. Populations evolve through the effects of mutation, selection, chance (genetic drift), and gene flow. Large outbreeding, sexually reproducing populations typically contain substantial genetic diversity, while small populations typically contain reduced levels. Genetic impacts of small population size on inbreeding, loss of genetic diversity, and population differentiation are determined by the genetically effective population size, which is usually much smaller than the number of individuals.

TERMS

Adaptive evolution, balancing selection, bottleneck, census population size, diversifying selection, effective population size, endangered, equilibrium, fitness, gene flow, genetic drift, heterozygosity, idealized population, identity by descent, inbreeding coefficient, lethal, mutation-selection balance, natural selection, neutral mutation, outbreeding, random mating

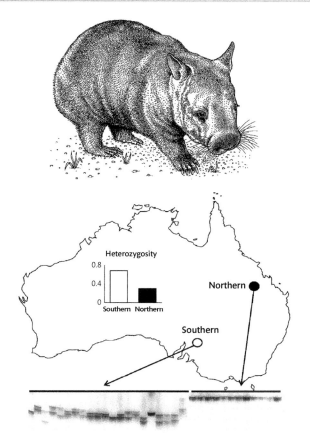

Low microsatellite genetic diversity in the critically endangered northern hairy-nosed wombat (*Lasiorhinus krefftii*), compared to its nearest relative, the southern hairy-nosed wombat (*Lasiorhinus latifrons*) in Australia (Frankham et al. 2017, p. 17, based on Taylor et al. 1994; Beheregaray et al. 2000). The autoradiograph below the map shows genotypes for a microsatellite in several individuals of the two species, with the northern species lacking variation.

A Practical Guide for Genetic Management of Fragmented Animal and Plant Populations. R. Frankham, J. D. Ballou, K. Ralls, M. D. B. Eldridge, M. R. Dudash, C. B. Fenster, R. C. Lacy & P. Sunnucks. Oxford University Press (2019). © R. Frankham, J. D. Ballou, K. Ralls, M. D. B. Eldridge, M. R. Dudash, C. B. Fenster, R. C. Lacy & P. Sunnucks 2019. DOI: 10.1093/oso/9780198783411.001.0001

Background

This book applies conservation and evolutionary genetics to the genetic management of fragmented animal and plant populations

Genetic management of fragmented populations is a component of conservation genetics, an applied discipline drawing upon evolutionary, population, molecular, and quantitative genetics, and genomics. This information is applied to conservation issues with particular emphasis on small populations (Fig. 2.1). These genetic issues are also influenced by demography, ecology, disease biology, etc., and may, in turn, influence them.

This chapter provides a brief introduction to the critical concepts of conservation and evolutionary genetics that we apply in the remainder of the book.

Fig. 2.1 Structure of conservation genetics and its relationship to other disciplines (Frankham et al. 2017, Fig. 2.1).

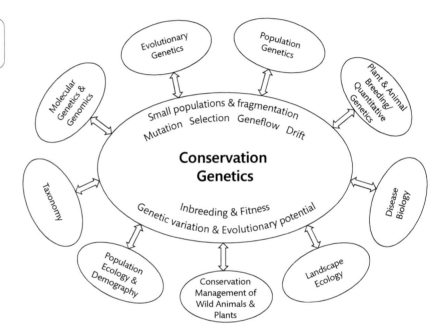

Populations evolve through the impacts of mutation, selection, chance, and gene flow

These factors have the following roles in the evolution of populations:

- mutation is the ultimate source of genetic diversity for evolution
- selection is the only force causing adaptive evolutionary changes in populations (Chapter 3)
- chance effects (genetic drift) deplete genetic diversity and lead to random genetic changes within and among populations over generations (Chapter 3)
- gene flow (often referred to as migration) moves genetic variants from one population fragment to another via individuals or gametes (Chapter 4).

These forces operate in the context of the species' reproductive system and mode of inheritance. In this book, we concentrate primarily on the sexually reproducing diploids that are usually the chief concern of conservation managers. The implications of alternative reproductive systems and modes of inheritance, such as self-fertilization, are considered more fully in Chapter 8 of Frankham et al. (2017).

Next, we elaborate on the operation of mutation, selection, and chance.

Mutation

We focus on the rate of occurrence of mutations and their effects on reproductive fitness.

Mutations occur at a low rate across the genome

New mutations arise at low rates that vary for different genetic characters. For example, mutations that affect physical features of house mice (*Mus musculus*), maize (*Zea mays*), and *Drosophila* fruit flies occur at a rate of about 1 in 100,000 genes each generation, whereas mutations affecting individual DNA nucleotides in primates are about 1,000 times less frequent (Pfeifer 2017).

Mutations can arise anywhere in the genome, including genes that encode proteins, in their regulatory regions, and in parts of the genome that are not known to have a function.

Most new mutations are harmful

Mutations have effects that are:

- harmful
- advantageous
- conditional on the environment
- conditional on other alleles in the same or other genes, or
- neutral (do not alter fitness).

In populations adapting to their environments, most new mutations are harmful, few are advantageous (Fig. 2.2), some are conditional, and the proportion that are neutral depends on the character being considered. Harmful mutations are pivotal to the effects of inbreeding on fitness (Chapter 3). Advantageous mutations and some conditional ones are utilized in adaptive evolution (Chapter 3). Geneticists use neutral mutations to characterize genetic diversity, genetic population structure, and rates of gene flow (Chapters 3, 4, and 7), and to build phylogenetic trees used in taxonomy (Chapter 6).

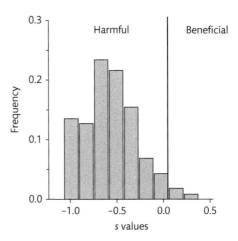

Fig. 2.2 Most new mutations are harmful in the environment to which the population is adapted. Distribution of mutational effects in a bacterial virus (Frankham et al. 2017, Fig. 2.2, based on Vale et al. 2012). The scale at the bottom is the selection coefficient (s), where values of −1.0 are lethal, < 0 harmful, 0 neutral, and > 0 beneficial.

Selection is the only force that produces adaptive evolutionary change

Selection increases the frequencies of genetic variants associated with greater fitness, leading to improved evolutionary adaption to the environment that a population inhabits. Selection may be:

- directional
- balancing, or
- diversifying.

Directional selection decreases the frequency of harmful alleles and increases the frequency of favorable ones (adaptive evolution).

Balancing selection actively maintains genetic diversity in populations when heterozygotes or rare alleles are favored.

Diversifying selection involves conditional alleles that are advantageous in some environments, but harmful in others, resulting in adaptively differentiated populations in diverse environments. Consequently, genotypes typically show better performance in their home site than in alternative environments.

Chance effects arise from random sampling of gametes during reproduction

Chance effects have important impacts on the evolution of small populations, but only minor impacts on large populations. Chance sampling of gametes to produce individuals in successive generations results in effects that are **random in direction**, in contrast to the directional effects of mutation, migration, and selection. Chance results in five related genetic consequences:

1. Random genetic drift: random changes in allele frequencies across generations within populations (Fig. 2.3)

18

2. Genetic differentiation among isolated populations. As drift in different isolated populations is random, they differentiate non-adaptively over generations (Fig. 2.3)
3. Loss of genetic diversity within isolated population fragments
4. Reduced heterozygosity across a group of population fragments, compared to that in a single connected population
5. Reduced effectiveness of selection in smaller than larger populations because chance plays a greater role in which alleles increase or decrease than does natural selection. In the experiment shown in Fig. 2.3 selection favored the wild-type allele in large populations, but was much less effective in small populations, including one replicate where the favored allele was lost by chance (asterisked).

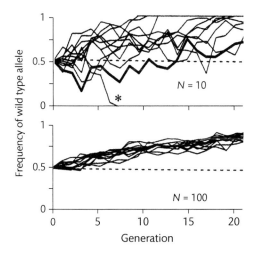

Fig. 2.3 Random genetic drift of the wild-type allele for body color in *Tribolium* flour beetles, along with selection favoring the wild-type allele. All 24 populations began with frequencies of 0.5 of the wild-type (red) and black alleles, and were maintained in isolation by random sampling of either 10 or 100 parents in each generation. *N* is the population size and the asterisk identifies a population that by chance became homozygous for the lower fitness black allele (Frankham et al. 2010, Fig. 8.2, after Rich et al. 1979).

Loss of genetic diversity is more rapid in smaller than larger populations

The average heterozygosity is reduced by drift due to population size restrictions (bottlenecks). In general, the reduction in heterozygosity due to a single generation of sampling is $1/(2N)$, where N is the population size. Thus, a bottleneck of one pair of parents is expected to reduce heterozygosity by ¼, and one with five pairs by 1/20. For example, single pair bottlenecks in populations of *Drosophila* fruit flies reduced heterozygosity for microsatellites (see Appendix 1) to 0.43 from 0.61 in the base population, and reduced allelic diversity to 2.49 from 7.0. Population bottlenecks have also been observed to reduce genetic diversity in threatened species, including Ethiopian wolves (*Canis simensis*), Mauritius kestrels (*Falco punctatus*), and Florida torreya trees (*Torreya taxifolia*).

We next consider what genetic markers are used in conservation genetics to measure and describe genetic diversity.

Florida torreya (USA)

How do we measure and describe genetic diversity?

Conservation genetics utilizes a range of genetic markers and characters, including DNA nucleotides, microsatellites, allozymes, morphological traits, and chromosomal variants to measure genetic diversity

For those unfamiliar with molecular genetic markers, such as microsatellites (short variable number DNA sequence repeats), SNPs (single nucleotide polymorphisms), and variation in DNA sequences, details are given in Appendix 1.

If conservation agencies are not able to assess genetic diversity in their populations of interest, they should be able to identify organizations that will be able to assist on a fee for service or collaborative basis. For example, the Australian Museum has a laboratory that identifies wildlife species and measures their genetic diversity for a fee, as do state-based and commercial service providers and research institutions around the world.

Multiple individual markers

Genetic diversity for individual genetic markers and genes in a population is described using the proportion of heterozygotes and the number of different alleles, and these are averaged over multiple genetic markers

Following genotyping of individuals from a population for a given marker, we summarize the variation based on the proportion of individuals that are heterozygous and the number of alleles for the marker, as illustrated in Box 2.1 for the Ethiopian wolf. Mean genetic diversity is characterized using observed heterozygosity (H_o), expected heterozygosity (H_e), and the average number of alleles per marker (A). Observed heterozygosity is simply the number of heterozygotes as a proportion of the total individuals examined. Expected heterozygosity is the proportion of heterozygotes expected from the observed allele frequencies under random mating, based on the Hardy–Weinberg equilibrium (Box 2.2). Expected heterozygosity has more desirable statistical properties than observed heterozygosity.

Box 2.1 Measuring and describing genetic diversity in the critically endangered Ethiopian wolf and in non-endangered wolves, coyotes, and dogs

(Frankham et al. 2017, Box 2.1, based on Gottelli et al. 1994; Anonymous 2003)

Ethiopian wolf

The Ethiopian wolf has a total population of < 400 individuals in only six isolated areas of Ethiopia. Genetic diversity for nine microsatellites in two populations, one on the Sanetti Plateau, where few domestic dogs were present, and the other in Web Valley, where dogs were abundant, is shown in the following table.

Taxon	A	H_o	H_e	n
Ethiopian wolf				
Sanetti	2.0	0.150	0.138	16
Web	2.8	0.313	0.271	23
Domestic dogs	6.4	0.516	0.679	35

Taxon	A	H_o	H_e	n
Gray wolf	4.5		0.620	18
Coyote	5.9		0.675	17

Three points are of note for conservation. First, the Ethiopian wolf populations have lower genetic diversity than the related, non-endangered gray wolves, coyotes (*Canis latrans*), and domestic dogs. Second, the relatively "pure" Sanetti population has lower genetic diversity than the Web Valley population that breeds with domestic dogs. Third, mating in the Ethiopian wolves is approximately random within populations, because observed and expected heterozygosities are similar in both populations.

Box 2.2 Hardy–Weinberg equilibrium genotype frequencies in large random mating populations

(Frankham et al. 2017, Box 2.2)

Genotypes at a diploid genetic marker attain Hardy–Weinberg equilibrium frequencies within one generation in a large random mating population, if there is no mutation, gene flow, drift, or selection. For example, if there are two alleles, A_1 and A_2, at frequencies of p and q (where $p + q = 1$), the gametes, zygotes, and their frequencies are given in the table below:

Allele and frequency		Ova	
		$p\ A_1$	$q\ A_2$
Sperm	$p\ A_1$	$p^2\ A_1A_1$	$pq\ A_1A_2$
	$q\ A_2$	$pq\ A_1A_2$	$q^2\ A_2A_2$

Thus, the zygotic frequencies for A_1A_1, A_1A_2, and A_2A_2 genotypes are p^2, $2pq$, and q^2, respectively. For example, if the two alleles have frequencies of 0.7 and 0.3, respectively, A_1A_1 has an expected frequency of $0.7^2 = 0.49$, A_1A_2 $2 \times 0.7 \times 0.3 = 0.42$, and A_2A_2 $0.3^2 = 0.09$ (and they sum to 1.0, as they must).

We now consider the joint operation of all the forces on the genetic diversity of populations, referred to as standing genetic diversity. Different components of this diversity predict the ability of populations to evolve, and the extent of harmful impacts of inbreeding on fitness.

Small populations have reduced levels of genetic diversity

Cheetah (Africa)

In general, large populations of species exhibiting approximately random mating (outbreeding) contain large stores of all forms of genetic diversity. Conversely, small populations have lower levels of genetic diversity.

Threatened species have on average 25–35% less microsatellite genetic diversity than related non-threatened ones, as found for example in cheetahs (*Acinonyx jubatus*), northern hairy-nosed wombats, Komodo dragons (*Varanus komodoensis*), and Zhiyuan fir trees (*Abies ziyuanensis*).

Populations harbor rare harmful alleles

Since mutation adds harmful alleles to populations but selection cannot remove them all immediately, particularly recessive ones, a balance is reached in which a typical gene has < 1% of harmful alleles. Because there are tens of thousands of genes in a genome, most individuals carry several harmful alleles hidden in heterozygotes.

Large populations of outbreeding species have a large store of rare harmful alleles (genetic load) that can be made homozygous by inbreeding (Chapter 3). For example, over 6,000 single gene disorders are known in humans (*Homo sapiens*), many due to rare harmful recessive alleles, such as those causing phenylketonuria and cystic fibrosis. Further, rare single gene chlorophyll deficiency mutations are found in many plant species.

Adaptive genetic variation

Komodo dragon (Indonesia)

Adaptive genetic variation is found for most fitness characters in most species. For example, genetic variation has been found for fecundity in house mice, rats (*Rattus* sp.), invertebrates, and maize, for fitness on altered diets in invertebrates and soils in plants, and for resistance to disease in humans, European rabbits (*Oryctolagus cuniculus*), and plants. Adaptive genetic variation is most readily observed when the environment changes, as many previously harmful conditional alleles become beneficial, and vice versa.

Since population size is a critical issue in conservation genetics, we next specify how we define the genetically effective population size and contrast it with the census population size (the number of individuals in the population).

Genetic impacts depend upon the genetically effective population size, rather than the census size

All evolutionary genetic processes depend upon the genetically effective population size (N_e), the size of an idealized population that results in the same genetic consequences as

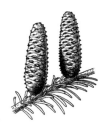

Zhiyuan fir (China)

experienced by a real population of interest (Box 2.3). N_e is based on the number of fertile adults that breed, but accounts for unequal contributions to the next generation from different breeders, variations in sex ratios of breeders, and fluctuation in population size across generations.

Box 2.3 Effective population size (N_e) and the idealized population
(after Frankham et al. 2010)

The effective size of a population (N_e) is the size of an idealized population that would lose genetic diversity, or become inbred, or drift, at the same rate as the actual population. For example, if a real population loses genetic diversity at the same rate as an ideal population of 50, then we say the real population has an effective size of 50, even if it contains 500 potentially reproductive adults.

The idealized population

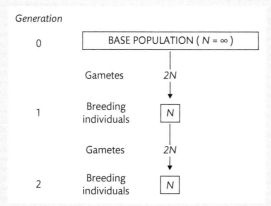

(Frankham et al. 2010, Fig. 8.8)

We begin by assuming a large (essentially infinite) random mating base population, from which we take a sample of N adults to form the idealized population (figure above). This population is maintained as a random mating, isolated population in succeeding generations. Alleles may be lost by chance and allele frequencies may fluctuate due to genetic drift. The simplifying conditions applied to the idealized population are:

- number of breeding individuals is constant in all generations
- generations are distinct and do not overlap
- no gene flow with other populations
- all adults are potential breeders
- all individuals are hermaphrodites (possess both female and male sex organs)

Continued ➤

- union of gametes is random, including the possibility of rare self-fertilizations
- no selection at any stage of the life cycle
- no mutation
- breeding individuals within populations contribute gametes equally to a pool, these unite at random to form diploid zygotes, resulting in random variation in numbers of offspring per adult.

Effective population sizes in natural populations are almost always less than census ones

Real populations deviate in structure from the assumptions of the idealized population by having variable numbers in successive generations, high variation in family sizes, unequal sex ratios, non-random mating, and overlapping generations. Consequently, their effective population sizes are usually much smaller than the number of potentially breeding adults. For example, the current human population contains ~ 4 billion potentially breeding adults, but its long-term effective size is only ~ 10,000 due to ancient bottlenecks in population size. The N_e of a population can be calculated for different time periods, ranging from the average N_e over thousands of generations (long-term N_e as given for humans above), to the N_e of a single generation. When we need to predict the future N_e of a small population for management purposes, we want to estimate the N_e for an intermediate time period, encompassing several recent generations (which we call multigenerational N_e). The long-term N_e is not useful for our purposes, because it will reflect events in the distant past and may be very different from the current N_e of a population. An estimate of N_e for a single recent generation is also unsuitable, because it does not incorporate fluctuations in population size. It is crucial that all relevant variables be included in estimates of multigenerational N_e for the many conservation applications where we deal with multiple generations.

For most species, N_e is unknown, so inferences are often based upon adult census sizes (N) and N_e/N ratios for those species where it is known. What are typical values for this ratio?

In unmanaged wild populations, multigenerational estimates of N_e/N ratios encompassing all relevant variables average 0.11–0.14, based on three meta-analyses. Although, N_e/N ratios vary widely with life history and breeding systems, N_e is almost always smaller than N in unmanaged populations. Consequently, adverse genetic effects in populations (loss of genetic diversity and inbreeding) typically occur much sooner and faster than expected from census population sizes.

> Multigenerational N_e/N ratios average 0.1–0.2, but vary widely across taxa

Obtaining estimates of N_e for your population

As N_e is used to make predictions later in the book, we provide guidelines in Box 2.4 on how to obtain a value for your population.

Box 2.4 How to obtain an estimate of effective population size for your population or species

(Wright 1931, 1969, Frankham 1995; Waples 1989; Palstra & Ruzzante 2008)

The simplest method for obtaining an N_e value is to use the current adult population size for your isolated population and multiply it by published information on N_e/N ratio for your species or a related one with a similar life history. As we use N_e to make multigenerational predictions, it is critical that the N_e/N ratio used encompasses the effects of fluctuation in population size over generations, variance in family sizes, and unequal sex ratios (Wright 1931, 1969). These three factors all cause N_e to be a smaller proportion of N, with fluctuations in population size over generations having the largest effect on average (Frankham 1995). The arithmetic mean of N values is used as this is available for more populations and species, and has been used more often than other measures.

For an average species N_e/N is ~ 1/8 (Frankham 1995; Palstra & Ruzzante 2008; Palstra & Fraser 2012). However, N_e is typically closer to N in smaller than larger populations (Pray et al. 1996; Hauser & Cavalho 2008).

If you are working on threatened species, the census population sizes under the IUCN's Red List criterion D for stable populations are $N \leq 50$ for critically endangered, ≤ 250 for endangered, and $\leq 1,000$ for vulnerable species (IUCN 2018). For an average species, these correspond to approximate effective sizes of ≤ 6, ≤ 31, and ≤ 125 for critically endangered, endangered, and vulnerable species, respectively.

However, values of N_e/N are much less than 1/8 for species with "fast" life cycles (high fecundity and short generation intervals), while species with "slow" life cycles (low fecundity and long generation intervals) typically show higher values (Frankham et al. 2010; Waples et al. 2013). For example, highly fecund species with high juvenile mortality such as marine fish, shrimp, copepods, and brown seaweed have estimated ratios of 10^{-3} to 10^{-8} (Coyer et al. 2008; Hauser & Carvalho 2008; Palstra & Ruzzante 2008; Zeller et al. 2008, but see Waples 2016). Consequently, it is best to use an empirical estimate of the multigenerational N_e/N ratio for your species, or a related species with a similar life history.

Given the diversity of single and multiple generation methods used to estimate N_e, the need to correctly match N_e and N, to account for different life histories and overlapping generations, the biases due to small sample size, etc. (Barker 2011), we have prepared online Appendix 2 with multigenerational N_e/N estimates for a range of taxa with a variety of life histories. These are empirical estimates for unmanaged populations in wild habitats. Note that individual estimates of N_e/N ratios often have large confidence intervals. If you are unsure of what value of N_e to use, we recommend that you do your assessments with values likely to represent the credible limits of N_e to determine whether both lead to similar management (as done in Box 7.4).

We next discuss the importance of inbreeding in conservation. It is one of the two major genetic factors affecting population persistence, along with loss of genetic diversity.

Inbreeding reduces fitness and increases extinction risk

Inbred offspring typically have reduced survival and reproduction compared to non-inbred ones (termed inbreeding depression). This is most severe in naturally outbreeding species, but is also found in most inbreeding species (Chapter 3).

Having introduced inbreeding, it is now time to define what it is and how we measure it.

Inbreeding is the production of offspring by individuals related by descent

Inbreeding encompasses the production of offspring by related individuals, whether they are closely related (e.g. self-fertilization, brother–sister, and parent–offspring matings), or more distantly related, such as members of the same small isolated population. The degree of inbreeding depends on how related the parents are through having ancestors in common.

The inbreeding coefficient is used to measure inbreeding

> The inbreeding coefficient F is the probability that two alleles of a genetic marker in an individual are identical because they were inherited from an ancestor of both parents (identical by descent)

Inbreeding coefficients can be assessed directly from pedigrees, either via direct probability calculations (Example 2.1) or by using computer software (e.g. PMx). However, for most wild populations, pedigrees are lacking and inbreeding must be estimated from genetic marker data (see Chapters 3, 7, and 8).

Example 2.1 Computing inbreeding coefficients for individuals resulting from self-fertilization and full-sib mating

(Frankham et al. 2010, Fig. 12.2)

To compute the inbreeding coefficient of individual X from pedigrees in the examples of self-fertilization and full-sib mating, we attribute distinct genotypes to the common ancestors (A_1A_2 and A_3A_4) and compute the probability that two alleles in X are identical by descent assuming Mendelian segregation. The lines indicate potential transmission pathways of the alleles from the offspring's common ancestor(s).

Self-fertilization

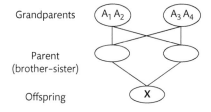

Parent $A_1 A_2$

Gametes $\frac{1}{2}A_1$ $\frac{1}{2}A_2$

Offspring   X

$$F_X = Pr(X = A_1A_1) + Pr(X = A_2A_2)$$
$$= \frac{1}{4} + \frac{1}{4} = \frac{1}{2}$$

Full-sib mating

Grandparents $A_1 A_2$ $A_3 A_4$

Parent
(brother–sister)

Offspring   X

$$F_X = Pr(X = A_1A_1 \text{ or } A_2A_2 \text{ or } A_3A_3 \text{ or } A_4A_4)$$
$$= \frac{1}{16} + \frac{1}{16} + \frac{1}{16} + \frac{1}{16} = \frac{1}{4}$$

Thus, the inbreeding coefficient of an individual resulting from self-fertilization is ½, and that for one resulting from full-sib mating is ¼, while for non-inbred individual it is zero.

Repeated inbreeding can produce even higher inbreeding coefficients: an F of 0.999 is reached after 10 generations of continuous self-fertilization, while nearly the same F of 0.986 occurs after 20 generations of brother–sister mating.

Because inbreeding increases homozygosity for all alleles, including harmful ones, it reduces reproduction and survival rates, a major issue in conservation genetics (Chapter 3).

Inbreeding is unavoidable in isolated random-mating populations

Even if a population is founded by completely unrelated individuals and mating is at random, in time all individuals will become related. Much of the inbreeding we need to manage in threatened species arises in this way. Over a single generation in a isolated random-mating population of size N_e, the increase in inbreeding is $1/(2N_e)$ (the same as the loss of heterozygosity). Thus, in a population of $N_e = 10$, inbreeding increases by $1/(2 \times 10) = 5\%$ in each generation. Inbreeding increases progressively across generations at a faster rate in smaller than larger populations, as illustrated in Fig. 2.4.

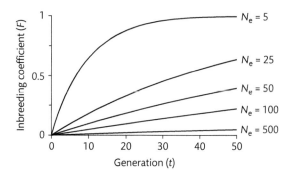

Fig. 2.4 Increase in inbreeding coefficient (*F*) with generations (*t*) in closed random mating populations of different sizes. N_e is effective population size (Frankham et al. 2010, Fig. 12.4).

In the succeeding chapters in this section we discuss genetic problems in small populations and how to reverse them. Chapter 3 deals with inbreeding depression, loss of genetic diversity, and reduced ability to evolve.

Summary

1. Genetic management of fragmented populations involves the application of evolutionary genetic theory and knowledge to alleviate problems due to inbreeding and loss of genetic diversity in small populations.
2. Populations evolve through the effects of mutation adding genetic diversity, selection favoring particular variants over others, chance (genetic drift) leading to random changes within and among populations over generations and loss of genetic diversity, and gene flow spreading genetic diversity among populations.
3. Large populations of outbreeding sexually reproducing species typically contain substantial genetic diversity, including neutral alleles, harmful alleles in mutation-selection balance, and advantageous alleles.
4. Small populations, including threatened species, typically contain less genetic diversity than large populations and non-threatened species.
5. Genetic impacts of small population size (inbreeding, loss of genetic diversity, and population differentiation) are determined by the effective population size, rather than the census number of individuals.
6. The effective size of a population is the size of an idealized population resulting in the same inbreeding, loss of neutral genetic diversity, or fluctuations in allele frequencies as the target population. The effective population size is nearly always much less than the census size.
7. Inbreeding is the production of offspring from mating between relatives.
8. Small isolated random-mating populations become inbred and lose genetic diversity at rates proportional to $1/(2N_e)$ per generation.

FURTHER READING

Allendorf et al. (2013) *Conservation and the Genetics of Populations*: A comprehensive textbook on conservation genetics, including the topics in Section I of this book.

Fenster et al. (2018) Clear, simple explanation of the consequences of small population size in conservation, with emphasis on fitness and the ability to evolve.

Frankham et al. (2004) *A Primer of Conservation Genetics*: A simple introduction to the topics in Section I of this book.

Frankham et al. (2010) *Introduction to Conservation Genetics*: A comprehensive treatment of the background to the topics in this book.

Frankham et al. (2017) *Genetic Management of Fragmented Animal and Plant Populations*: Chapter 2 has a more detailed treatment of topics in this chapter.

SOFTWARE

PMx: a program for computing inbreeding coefficients and other measures from pedigrees (Lacy et al. 2012; Ballou et al. 2018). https://scti.tools/pmx

Inbreeding and loss of genetic diversity increase extinction risk

Inbreeding and loss of genetic diversity are unavoidable in small isolated populations, resulting in reduced fitness, compromised ability to evolve, and elevated extinction risks. Inbreeding depression is nearly universal in sexually reproducing diploid organisms, and its effects are typically very large for total fitness, especially in species that naturally outbreed. Inbreeding reduces fitness due to increased homozygosity for harmful alleles and at genes exhibiting heterozygote advantage. Species face ubiquitous environmental change and must adapt or they will go extinct. Loss of genetic diversity reduces the ability of populations and species to evolve. Threatened species typically have compromised ability to evolve due to lower genetic diversity, lower reproductive rates, and smaller effective population sizes than non-threatened species.

TERMS

Adaptive evolution, catastrophes, cline, directional selection, extinction vortex, heterozygote advantage, lethal equivalents, purging, quantitative character, relative fitness

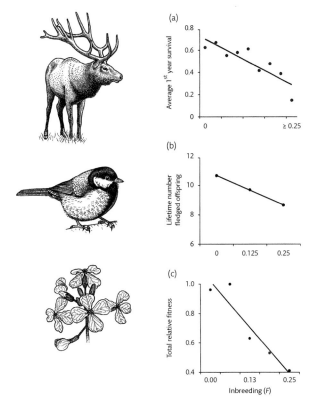

(a)

(b)

(c)

Inbreeding depression for fitness in the wild for (a) red deer (*Cervus elaphus*) (Scotland), (b) great tits (*Parus major*) (UK), and (c) wild radish (*Raphanus sativus*) (USA) (Nason & Ellstrand 1995; Szulkin et al. 2007; Walling et al. 2011). In all three, fitness declines progressively with increasing inbreeding.

A Practical Guide for Genetic Management of Fragmented Animal and Plant Populations. R. Frankham, J. D. Ballou, K. Ralls, M. D. B. Eldridge, M. R. Dudash, C. B. Fenster, R. C. Lacy & P. Sunnucks. Oxford University Press (2019). © R. Frankham, J. D. Ballou, K. Ralls, M. D. B. Eldridge, M. R. Dudash, C. B. Fenster, R. C. Lacy & P. Sunnucks 2019. DOI: 10.1093/oso/9780198783411.001.0001

Genetic erosion is unavoidable in small isolated populations

Adverse genetic effects typically occur in small isolated populations of outbreeding species due to inbreeding and loss of genetic diversity, leading to elevated extinct risks. In what follows, we describe the harmful effects on fitness and ability to evolve, as a prelude to later chapters where we describe management to prevent or reverse the harm.

Inbreeding reduces fitness

Inbreeding reduces reproductive fitness in naturally outbreeding species, and to a lesser extent in naturally inbreeding species. For example, in the 1870s Darwin found that self-fertilization reduced seed production by an average of 41%, compared to cross-fertilization, based on studies of 23 species of plants. Subsequently, inbreeding has also been shown to reduce fitness in laboratory, domesticated and wild animals and plants, and humans.

While there was initially skepticism about whether inbreeding was harmful in wildlife species, it reduced juvenile survival in 41 of 44 wild mammal species in captivity compared to outbred individuals (Fig. 3.1). For example, mortality in the pygmy hippopotamus (*Choeropsis liberiensis*) was 55% in inbred offspring versus 25% in outbred ones.

Fig. 3.1 Inbreeding depression for juvenile survival in 44 captive mammal populations (Ralls & Ballou 1983). Juvenile mortality of outbred individuals is plotted against that of inbred individuals from the same populations; inbreeding is harmful below the dotted line (after Frankham et al. 2010, Fig. 12.1).

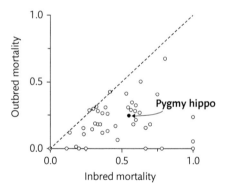

Pygmy hippopotamus (West Africa)

Inbreeding depression also occurs in wild species in natural habitats, and its impacts are typically more severe than in captivity. For example, inbreeding depression was detected in 90% of 157 data sets from a broad array of animal and plant taxa in natural environments, and the average impact was seven-fold greater than in captivity.

The harmful impacts of inbreeding elevate extinction risks, as illustrated in Box 3.1. We elaborate on this later in the chapter.

Box 3.1   Inbreeding increased extinction risk in butterfly populations in Finland

(Frankham et al. 2017, Box 3.1, based on Saccheri et al. 1998; Nieminen et al. 2001)

Levels of molecular heterozygosity were determined in 42 Glanville fritillary butterfly populations in Finland in 1995, and their fate recorded the following year: 35 populations persisted and seven went extinct by autumn 1996. Extinction rates were higher for populations with lower heterozygosity (more inbred) (graph, below), even after accounting for the effects of demographic and environmental variables known to affect extinction risks. Inbreeding explained 26% of the variation in extinction rate.

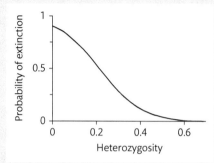

(Frankham et al. 2010, p. 24)

The causal link between inbreeding and extinction risk was confirmed by placing butterflies that were outbred, or inbred by brother–sister mating, in the field: all six inbred populations went extinct within one generation, while four of six outbred populations persisted.

Inbreeding reduces fitness because it increases homozygosity for harmful recessives and for genes exhibiting heterozygote advantage

Inbreeding depression arises because all naturally outbreeding populations contain harmful alleles (Chapter 2) and inbreeding, by increasing homozygosity, exposes harmful recessive alleles, and reduces heterozygosity at genes where heterozygotes have higher fitness than homozygotes (heterozygote advantage).

Factors affecting the magnitude of inbreeding depression

The magnitude of inbreeding depression depends upon the:

- number of harmful alleles segregating in the genome
- proportion of harmful alleles that are recessive
- increase in the inbreeding coefficient (ΔF)

- increase in homozygosity for harmful alleles
- environment in which individuals are located.

Inbreeding depression has the following characteristics:

- is ubiquitous for fitness in outbreeding diploid or polyploid species
- is greater for fitness than for traits with limited relationships to fitness (peripheral traits)
- occurs for all components of fitness
- accumulates over the life cycle
- is worse in more stressful environments
- varies across lineages, families, and species
- can be reduced by natural selection under some circumstances (purging)
- occurs in small isolated random mating populations.

We now elaborate on each of these.

Inbreeding depression for fitness is ubiquitous

Inbreeding depression for fitness is found in essentially all naturally outbreeding populations with adequate data, including mammals, birds, reptiles, fish, invertebrates, and plants (e.g. African lions, Florida panthers, golden lion tamarins, gray wolves, Mexican wolves, white-footed mice [*Peromyscus leucopus*], common shrews [*Sorex araneus*], Soay sheep [*Ovis aries*]), greater prairie chickens, Mexican jays [*Aphelocoma wollweberi*], song sparrows [*Melospiza melodia*], red-cockaded woodpeckers [*Picoides borealis*], great reed warblers [*Acrocephalus arundinaceus*], European adders, Atlantic salmon [*Salmo salar*], desert topminnows, rainbow trout [*Oncorhynchus mykiss*], land snails [*Arianta arbustorum*], insects, and plants [button wrinklewort daisies, maize, rose pink, and wild radish]).

Inbreeding depression might not be detectable among individuals within populations lacking genetic diversity for fitness due to a long history of small population size and inbreeding, but all individuals in such populations are likely to be experiencing inbreeding depression from previous inbreeding.

Thus, the assumption for an unstudied outbreeding species must be that more inbred individuals within the population will suffer reduced fitness if inbred, and it will already be suffering inbreeding depression if it has previously been inbred.

Inbreeding depression increases with the amount of inbreeding

The mean of most quantitative traits declines with increasing inbreeding (F) in an approximately linear fashion (chapter frontispiece; Fig. 3.2).

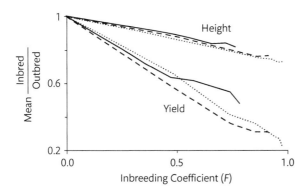

Fig. 3.2 Approximately linear declines of grain yield (a fitness trait) and height (a peripheral trait) in populations of maize inbred by selfing (dotted and dashed lines) or full-sib mating (solid lines) (Frankham et al. 2010, Fig. 13.4, after Falconer & Mackay 1996).

Inbreeding depression occurs for all aspects of reproductive fitness

Inbreeding is harmful for all aspects of reproductive fitness in animals and plants, such as sperm production, sperm quality, mating ability, female fecundity, juvenile survival, mothering ability, age at sexual maturity, predator avoidance, cancer risk, adult survival and longevity in animals, and related components in plants (Ujvari et al. 2018).

Not all studies report inbreeding depression for all characters studied, but virtually all show it for most components of reproductive fitness. Inbreeding depression is typically greater for fitness traits such as yield than for traits peripherally related to fitness, such as height (Fig. 3.2).

Inbreeding depression accumulates across life cycle stages

As inbreeding depression occurs for all aspects of fitness, it accumulates over the life cycle, as observed in animals such as the great tit (Fig. 3.3) and takahe birds, red deer, and many plant species. **Consequently, information on total fitness is required to capture the full impacts of inbreeding.** Maternal inbreeding also reduces offspring fitness: that is, offspring with an inbred mother are likely to be less fit than if they had a non-inbred mother. Consequently, both maternal and individual inbreeding effects must be determined to encompass the full effects of inbreeding on fitness, as for example by extending studies to grandoffspring recruitments (Fig. 3.3).

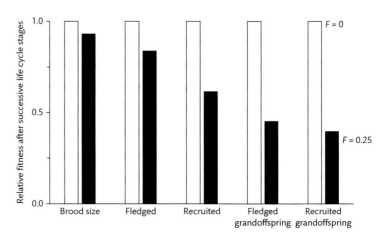

Fig. 3.3 Cumulative inbreeding depression across the life cycle of the great tit (Frankham et al. 2017, Fig. 3.6, after Szulkin et al. 2007). Black bars represent the relative fitness of inbred offspring ($F = 0.25$), compared to outbred offspring (white bars, $F = 0$) after inclusion of successive life history traits.

Rose pink plant (USA)

Inbreeding depression is greater in more stressful environments

Inbreeding depression is on average 69% greater in stressful than in benign environments and increases linearly with the stressfulness of environments. For example, inbreeding depression in the rose pink plant was 29% greater for total fitness in the field than in the greenhouse and in the great tit was 2.5 times greater in poor than in good environments.

Harmful impacts of inbreeding vary among individuals, families, populations, and species

Since inbreeding depression depends on the frequency of homozygotes for harmful alleles, it exhibits variability. For example, such variation in inbreeding depression among lineages has been reported within many species, including plants, old-field mice (*Peromyscus polionotus*), dairy cattle (*Bos taurus*), *Drosophila* fruit flies, and flour beetles.

Natural selection can reduce inbreeding depression

Selection can reduce the frequency of harmful recessive alleles (purging), but it cannot reduce inbreeding depression due to heterozygote advantage. Partial purging of the load of harmful alleles has been documented in many species, including plants, mice, gorillas (*Gorilla gorilla*), birds, and fruit flies. As many populations of conservation concern have small population sizes, where natural selection is inefficient, purging effects are often small in conservation contexts. Further, even if purging reduces inbreeding depression for a population in one environment, it may not lead to benefits in other environments, because often the deleterious effects of alleles are environmentally dependent.

Inbreeding depression occurs in small isolated random mating populations

Most inbreeding of conservation concern in naturally outbreeding species occurs gradually in small isolated populations. The inbreeding in such populations increases by a proportion of $1/(2N_e)$ per generation. For example, an inbreeding coefficient of 0.25 is reached after 57 generations in a population with an N_e of 100 (Example 3.1).

Example 3.1 Inbreeding level in an isolated finite population

If we start with a non-inbred population ($F = 0$), the inbreeding coefficient after 57 generations (F_t) in an isolated random mating population with an effective size of 100 is computed as follows:

$$F_t = 1 - (1 - \frac{1}{2N_e})^t = 1 - (1 - \frac{1}{2 \times 100})^{57} = 0.25$$

This is the same level of inbreeding as that in the progeny of brother–sister or parent–offspring matings.

Slow inbreeding reduces population fitness in small populations, as observed in black-footed rock-wallabies, greater prairie chickens, European adders (Box 3.2), topminnow fish, *Drosophila* fruit flies, house flies (*Musca domestica*), butterflies, and a variety of plants.

Box 3.2 Inbreeding depression in a small isolated population of European adders in Sweden

(Frankham et al. 2017, Box 3.2, based on Madsen et al. 1996, 2004)

In Sweden, a small isolated population of adders (< 40 individuals), separated from the main distribution of the snake for at least a century, has low molecular genetic diversity, and is inbred relative to the main population.

The small population exhibited inbreeding depression for litter size and proportion of abnormal offspring, compared to the larger population. Different environmental conditions were ruled out as an explanation for the abnormal offspring, because the progeny of matings between an introduced male from the large population and females from the small population had far fewer abnormalities.

European adder

Inbreeding effects on total fitness are typically extremely harmful

While inbreeding depression for individual fitness components may be modest, effects on total fitness in the wild are typically very large in naturally outbreeding species (Table 3.1). Further, half of those values are underestimates as they do not include the harmful impacts of maternal inbreeding.

Table 3.1 Inbreeding depression (ID) for total fitness in wild species of animals and plants due to a 25% increase in the inbreeding coefficient. Haploid lethal equivalents (L.E.), described later in the chapter, are also reported (after Frankham et al. 2017, Table 3.2).

Common name	Genus and species	ID %	L.E.
Red deer	*Cervus elaphus*	99	18.7
Collared flycatcher	*Ficedula albicollis*	94[a]	7.5[a]
Great tit	*Parus major*	55	3.2
Song sparrow	*Melospiza melodia*	79	6.2
Takahe	*Porphyrio hochstetteri*	88	8.0
Deerhorn clarkia	*Clarkia pulchella*	100[a]	39.2[a]
Rose pink plant	*Sabatia angularis*	38[a]	1.9[a]
Wild radish	*Raphanus sativus*	56[a]	3.3[a]

[a] Maternal inbreeding contribution not included, or not fully included.

Inbreeding increases extinction risks

Evidence from captive and wild populations, and computer modeling show that inbreeding increases extinction risk in outbreeding populations. For example, sustained inbreeding has been shown to increase extinction rates in populations of many species, including *Drosophila* fruit flies, house flies, mice, Japanese quail (*Coturnix coturnix japonica*), maize, Italian ryegrass (*Lolium multiflorum*), and *Mimulus guttatus* plants. Even slow inbreeding due to small population size (N_e = 10 or 20, sizes typical of endangered species) causes extinctions, albeit at a rate lower than with full-sib mating (Fig. 3.4).

Fig. 3.4 Proportion of populations going extinct rises with inbreeding (*F*) in populations of *Drosophila* fruit flies maintained using full-sib mating or effective population sizes of 10 and 20 with random mating (Frankham et al. 2010, Fig. 13.5, after Reed et al. 2003). Causes of extinction other than inbreeding depression can be ruled out in these populations.

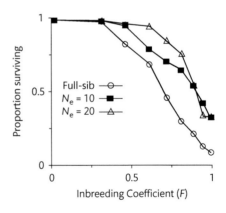

Inbreeding elevates extinction risks in wild populations

Although wild populations experience multiple threats, direct empirical evidence and computer projections demonstrate that inbreeding elevates their risk of extinction. First, inbreeding was a significant predictor of extinction risk for butterfly populations in Finland after the effects of all other ecological and demographic variables had been removed (Box 3.1). Similarly, experimental populations of the deerhorn clarkia plant with higher inbreeding ($F = 0.08$) exhibited a 69% extinction rate over three generations in the wild, while populations with lower inbreeding ($F = 0.04$) showed only a 25% extinction rate. Thus, a small difference in inbreeding translated into a large difference in population extinction rate.

Second, almost all computer projections for a range of outbreeding birds, mammals, reptiles, amphibians, and plants yielded substantial increases in extinction risk when the effects of inbreeding were included, as compared to when they were excluded (Fig. 3.5).

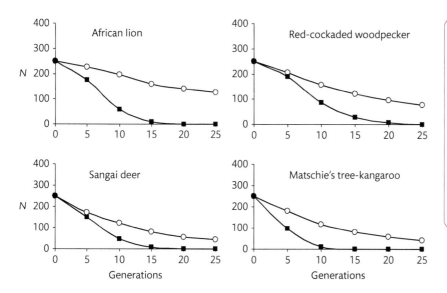

Fig. 3.5 Inbreeding is projected to substantially increase extinction risk in wild populations. Mean census population sizes (*N*) of 1,000 computer projections of population sizes for threatened wild populations of African lion, red-cockaded woodpecker, Matschie's tree-kangaroo (*Dendrolagus matschiei*), and Sangai deer (*Rucervus eldii eldii*) when the harmful effects of inbreeding on fitness are included (■) or excluded (○) in addition to all other variable and deterministic threats (Frankham et al. 2010, Fig. 2.4, after O'Grady et al. 2006).

Inbreeding interacts with non-genetic threats in an extinction vortex

Genetic and demographic factors interact in a negative feedback loop in small populations that accelerates the rate of decline in an "**extinction vortex**" (Fig. 3.6). If populations become small for any reason (human impacts, demographic or environmental variability, or catastrophes), they become more inbred and less demographically stable, further reducing population size and increasing inbreeding further. Such extinction vortices have been observed in several populations of vertebrates and plants and are expected to be widespread.

Fig. 3.6 The extinction vortex describes the likely interactions among human impacts, inbreeding, loss of genetic diversity, and demographic instability resulting in a downward spiral towards extinction (after Frankham et al. 2010, Fig. 2.2). N is census population size.

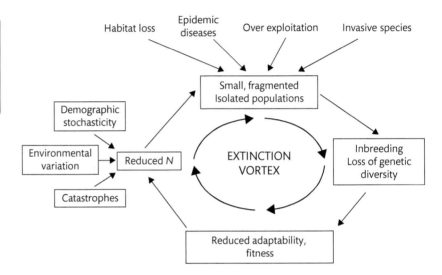

Comparisons of inbred and outbred individuals in the same environment are required to detect inbreeding depression

As survival and reproduction are strongly influenced by environmental conditions, tests for inbreeding depression require comparisons of the fitness of inbred and non-inbred individuals under the same environmental conditions. For example, the effects of inbreeding for white-footed mice in the wild were documented by simultaneously releasing inbred and outbred offspring of the same age into the wild and following their subsequent survival.

Inbreeding depression is quantified as the decline in trait mean due to an increase in inbreeding

The most common measures of inbreeding depression in plants and animals are proportionate decline in mean of a quantitative trait due to inbreeding, delta (δ), and the number of lethal equivalents, as described below.

Proportional decline in fitness due to inbreeding

Delta provides a point estimate of inbreeding depression from information on the impact of one level of inbreeding versus non-inbred individuals, as illustrated in Example 3.2. Delta should be accompanied by specification of the levels of inbreeding being compared. In plants (where it is widely used) it typically involves comparing selfed ($F = 0.5$) and outcrossed progeny ($F = 0$). The estimates of inbreeding depression due to brother–sister mating ($F = 0.25$) in Table 3.1 are presented in this form.

Example 3.2 Inbreeding depression for total fitness in rose pink plants

(Frankham et al. 2017, Example 3.1)

Total fitness in the field of propagules resulting from selfing and outcrossing were 0.25 and 0.66, respectively (Dudash 1990). Thus, the inbreeding depression (δ) is:

$$\delta = 1 - \frac{inbred\ fitness}{outbred\ fitness} = 1 - \frac{0.25}{0.66} = 0.62$$

This fitness decline of 62% is due to the inbreeding coefficient being 0.5 higher in the inbred than outbred offspring.

Lethal equivalents

A lethal equivalent refers to a group of harmful alleles that would cause death or sterility if homozygous, e.g. one lethal or sterility allele, two alleles that each cause a 50% probability of death or sterility, etc. Lethal equivalents can be estimated from the slope (B) of the regression of natural logarithm (ln) of relative fitness on level of inbreeding, using data on fitness of individuals with different levels of inbreeding measured in the same environment. B measures the additional genetic damage that would be expressed in a complete homozygote ($F = 1$) and is the number of haploid lethal equivalents. Inbreeding levels can be obtained from pedigrees or from genetic marker data, with the latter generally being required for wild populations (Chapter 7).

For example, in golden lion tamarins the slope of the regression of ln (7-day survival) on F is $B = -1.72$ (Fig. 3.7), so the population contains 1.72 haploid lethal equivalents for early survival.

> Lethal equivalents are used for quantifying the extent of inbreeding depression for survival or relative fitness. They are estimated from the slope of the decline in mean with increasing inbreeding

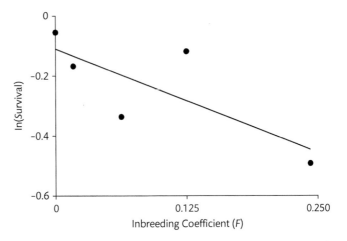

Fig. 3.7 Relationships between survival to 7 days and inbreeding coefficient in golden lion tamarins in natural habitats in Brazil (Frankham et al. 2017, Fig. 3.12, based on Dietz et al. 2000). The natural logarithm of survival [ln (Survival)] is plotted against the inbreeding coefficient (*F*) along with the best fitting linear regression line: ln$S$ = – 0.057 – 1.717 F (P = 0.01).

Estimates of haploid lethal equivalents for total fitness in wild populations of out-breeding species range from 1.9 to 39.2, with a median of 6.85 (Table 3.1). Sampling variation explains much of the wide range, but there are also likely true differences among populations and species, and different methods can yield different estimates (Nietlisbach et al. 2018).

We now consider how genetic factors affect the ability to evolve.

Evolutionary changes may allow populations to cope with conditions that they could not previously tolerate

Species must cope with many changes over time in physical and biotic environments (e.g. pests, parasites, diseases, predators, and competitors), including those wrought by human activities. Adaptation to changed environments may take the form of immediate morphological, physiological, or behavioral modifications, or **evolutionary (genetic) adaptation** where natural selection alters the genetic composition of populations over generations. There is a limit to non-genetic adaptation, as it is typically confined to a single generation, and if environmental changes are greater than individuals can tolerate, then the species will go extinct.

Conversely, adaptive evolutionary change through natural selection continuing over generations may allow populations to survive and reproduce under conditions more extreme than any individual could originally tolerate.

In the remainder of this chapter, we provide evidence on the ubiquity and magnitude of adaptive evolutionary change, and consider the factors controlling it.

Evolutionary adaptation is ubiquitous in species with genetic diversity

Adaptive evolutionary changes allow species to inhabit almost every niche on Earth, from Mount Everest to deep ocean trenches, arctic saline pools at –23° C to boiling thermal springs and deep-sea vents, and from oceans and freshwater to deserts. Evolutionary changes are documented in animal morphology, behavior, color, prey size, body size, life history attributes, disease resistance, predator avoidance, tolerance to pollutants, biocide resistance, etc. (Voyles et al. 2018). Adaptive evolutionary changes in plants include those to soil conditions, water stress, flooding, light regimes, exposure to wind, resistance to disease, grazing, air pollution, and herbicides.

Adaptive evolutionary changes may be rapid and large

Rapid evolutionary changes are documented in many invertebrates, vertebrates, and plants. For example, they have occurred in beak and body dimensions in Darwin's medium ground finch (*Geospiza fortis*), migration patterns and rates in birds and plants, life history strategies in Trinidadian guppies (*Poecilia reticulata*), and flowering time and heavy metal tolerance in many species of plants. Notably, many species rapidly evolved resistance to biocontrol agents.

What determines the rate and magnitude of change over generations?

To evolve, species need genetic diversity, reproductive excess, and selection

Cumulative genetic adaptation over multiple generations depends upon:

- genetic diversity for the trait
- intensity of selection
- effective population size
- number of generations.

Naturally outbreeding species with large populations normally possess genetic diversity among individuals (Chapter 2), allowing evolutionary adaptation through natural selection.

We now deal in more detail with each of the factors affecting adaptive evolution.

The ability to evolve depends on genetic diversity for fitness, a quantitative character

Most adaptive evolution is for fitness, a quantitative character, typically involving alleles at many genes. In experimental studies in the laboratory and field, selection response increases with level of genetic diversity, as predicted (Fig. 3.8).

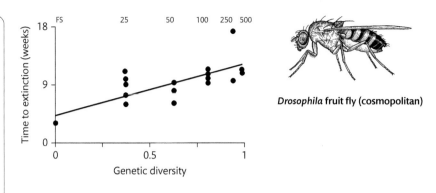

Fig. 3.8 Ability to evolve in small *Drosophila* fruit fly populations increases with genetic diversity (Frankham et al. 2010, p.238). Replicate populations were maintained at different effective sizes (numbers at top of figure) for 50 generations to generate the differences in genetic diversity. FS refers to populations full-sib mated for 35 generations. Equal numbers were used to establish large cage populations, which were subjected to increasing concentrations of NaCl until they went extinct.

Drosophila **fruit fly (cosmopolitan)**

Adaptive genetic diversity can be inferred indirectly via genomic diversity

In conservation contexts we often wish to know adaptive genetic diversity for a changed future environment. The most practical approach to determining adaptive genetic variation in most conservation contexts is to infer it from measures of genome-wide diversity for SNPs or microsatellites.

Loss of genetic diversity is unavoidable in small isolated populations

All isolated populations lose genetic diversity due to chance sampling of alleles during reproduction, with the rate being faster in smaller than larger populations (Chapter 2). Over a single generation in a random mating population of size N_e, the loss of heterozygosity is $1/(2N_e)$. For example, an isolated population of size $N_e = 50$ loses 1% of its genetic diversity each generation. Over multiple generations, loss of genetic diversity for neutral genetic markers continues to decline at an approximately exponential rate that is greater in smaller than larger populations (Fig. 3.9). An example of the prediction of such values is given in Example 3.3.

Fig. 3.9 Predicted decline in heterozygosity over generations in populations with different effective sizes (N_e) (Frankham et al. 2010, Fig. 11.1, after Foose 1986).

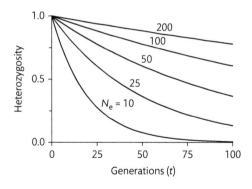

44

Example 3.3 Predicting loss of genetic diversity in isolated random mating populations over generations

The proportion of original neutral heterozygosity retained after t generations (H_t/H_0) is predicted by the following equation, where H_t is Hardy–Weinberg expected heterozygosity at generation t, and H_0 initial heterozygosity. For example, we compute the loss of heterozygosity after 20 generations in populations with effective sizes of 25 (relevant to many vulnerable species), as follows:

$$\frac{H_t}{H_0} = (1 - \frac{1}{2N_e})^t = (1 - \frac{1}{2 \times 25})^{20} = (\frac{49}{50})^{20} = 0.668$$

Thus, a population with N_e of 25 (~ 200 adults for a typical species) is predicted to lose 33% of its heterozygosity over 20 generations.

Magnitude of selection

The dependence of evolutionary change on the intensity of selection has been demonstrated in many artificial selection studies. In wild habitats, adaptation is more likely with greater habitat differences, an indicator of selection intensity.

Threatened species typically have reduced ability to evolve due to the combined effects of low genetic diversity, reduced magnitude of selection, and low N_e

Many threatened species have reduced genetic diversity for fitness characters, because they have on average 25–35% lower genetic diversity than related non-threatened species. Further, they probably have reduced fitness (due to inbreeding depression), compared to related, non-threatened species, and with fewer progeny produced per generation, the magnitude of selection is reduced. While we are not aware of any empirical data on evolutionary rates in threatened species, we apply what we know from many experimental evolution studies.

We assess the potential impacts of the above factors on the ability of the small Isle Royale population of gray wolves to evolve. At foundation, the island wolves had an estimated ~ 56% poorer ability to evolve over five generations than the large mainland population due to the combined effects of an initial 19% reduction in heterozygosity due to a bottleneck at foundation (one female and two males), a 35% reduction in the magnitude of selection due to inbreeding depression in litter size, a 99.5% lower N_e (5 versus 1,000), and a consequent 40-times faster loss of genetic diversity per generation (10% versus 0.25%).

The arguments above emphasize the importance of expanding the population sizes of threatened species to increase their effective sizes, thereby minimizing loss of genetic diversity and inbreeding, and improving their ability to evolve in response to environmental change.

Adaptation in heterogeneous environments

Diverse selective forces in different locations and no gene flow result in genetically differentiated populations

Eucalypt trees (Australia)

So far, we have dealt with consistent selective forces within a species. However, selection often differs among localities, especially if they have dissimilar environments. When there is little or no gene flow between populations in different localities, such diversifying selection leads to different genetic adaptations, provided the population fragments contain genetic diversity for fitness (Chapters 5, 6, 7, and 9). Given sufficient generations, this differential adaptation can lead to outbreeding depression or even reproductive isolation if the populations are crossed. Gene flow between population fragments reduces the extent of adaptive differentiation.

Where there is a selection gradient, as with temperature at different altitudes or latitudes, or day length with latitude, selection and migration may lead to a gradation in genetic adaptation across the landscape (termed a cline). For example, there are clines due to a balance between selection and gene flow in leaf waxiness with elevation in several species of eucalypt trees in Tasmania, Australia, and for cryptic coloration in peppered moths (*Biston betularia*) across gradients from polluted to unpolluted areas in England.

Inbreeding depression and compromised ability to adapt can often be reversed by gene flow

Since most harmful alleles involved in inbreeding depression are recessive, or exhibit heterozygote advantage, crossing inbred populations to unrelated populations within species can usually reverse the harmful fitness effects of inbreeding. Low genetic diversity and compromised ability to evolve in small fragmented populations also can be reversed by crossing with another population, as detailed in Chapter 5.

In the next chapter we describe the harmful genetic effects of population fragmentation.

Summary

1. Inbreeding and loss of genetic diversity are unavoidable in small isolated populations.
2. Inbreeding reduces reproductive fitness and increases the risk of extinction in essentially all well-studied populations of naturally outbreeding species. Thus, populations of outbreeding species that are inbred or have low genetic diversity should be presumed to be experiencing inbreeding depression and managed appropriately without waiting for specific evidence of inbreeding depression.
3. Inbreeding depression is due to increased homozygosity for harmful recessive alleles and at genes exhibiting heterozygote advantage.
4. Inbreeding depression is typically greater (a) in stressful than in benign conditions, (b) for total fitness than for individual fitness components, and (c) in naturally outbreeding than inbreeding species.
5. Natural selection may remove some alleles that cause inbreeding depression (purging), especially following inbreeding or population bottlenecks, but purging has

limited effects in small populations of conservation concern, especially when environments are changing.

6. The ability of populations to undergo adaptive evolution depends upon the amount of adaptive genetic diversity, intensity of selection, effective population size, and number of generations.

7. Loss of genetic diversity in small populations reduces their ability to evolve to cope with environmental change, thus increasing their extinction risk.

8. Increasing the population size of an endangered species as rapidly as possible minimizes inbreeding, loss of genetic diversity, and reduction in the population's ability to evolve.

FURTHER READING

Charlesworth & Willis (2009) Review on the genetics of inbreeding depression.
Frankham et al. (2017) *Genetic Management of Fragmented Animal and Plant Populations*: Chapters 3 and 4 have more detailed treatments of topics in this chapter.
Li et al. (2014) Genome sequencing study reporting that threatened bird species had lower heterozygosity, and more harmful alleles than non-threatened ones.
Thompson (2013) *Relentless Evolution*: Excellent recent review of evolution, documenting its pervasive nature and its frequent rapidity.

SOFTWARE

COANCESTRY: a program for simulating, estimating, and analyzing relatedness and inbreeding coefficients (Wang 2011). http://www.zsl.org/science/software/coancestry
PMx: a program for demographic and genetic management, including computing inbreeding coefficients from pedigrees (Lacy et al. 2012; Ballou et al. 2018). https://scti.tools/pmx

Population fragmentation causes inadequate gene flow and increases extinction risk

Most species now have fragmented distributions, often with adverse genetic consequences. The genetic impacts of population fragmentation depend critically upon gene flow among fragments and their effective sizes. Isolated population fragments deteriorate genetically over generations, resulting in greater inbreeding, increased loss of genetic diversity, decreased likelihood of evolutionary adaptation, and elevated extinction risk, when compared to a single population of the same total size. Even low levels of gene flow among population fragments are sufficient to prevent the harmful genetic effects of fragmentation. Detecting geographic clusters of individuals with similar genotypes is used to determine the number of genetically isolated population fragments and their boundaries, providing basic information for genetic management, especially for restoring gene flow.

TERMS

Coancestry, connectivity, F_{ST}, F statistics, habitat matrix, isolation by distance, isolation by environment, kinship, metapopulation, source–sink, Wahlund effect

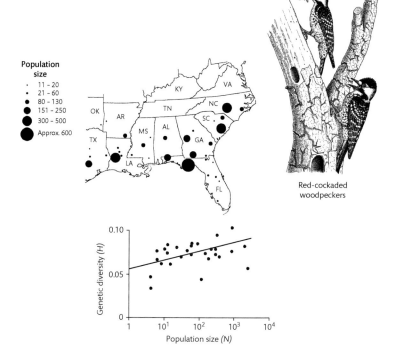

Red-cockaded woodpeckers

Population fragmentation in the red-cockaded woodpecker in southeastern USA (Kulhavy et al. 1995). The species had an essentially continuous distribution prior to European settlement, but is now highly fragmented (map after James 1995). Genetic diversity is on average less in smaller than larger populations (Frankham et al. 2010, p. 311, graph after Meffe & Carroll 1997).

A Practical Guide for Genetic Management of Fragmented Animal and Plant Populations. R. Frankham, J. D. Ballou, K. Ralls, M. D. B. Eldridge, M. R. Dudash, C. B. Fenster, R. C. Lacy & P. Sunnucks. Oxford University Press (2019). © R. Frankham, J. D. Ballou, K. Ralls, M. D. B. Eldridge, M. R. Dudash, C. B. Fenster, R. C. Lacy & P. Sunnucks 2019. DOI: 10.1093/oso/9780198783411.001.0001

Most species have fragmented distributions, many with adverse genetic effects on population persistence

Human-induced habitat loss and fragmentation is one of the primary causes of biodiversity loss. There are innumerable examples of habitat fragmentation throughout the world, including the highly speciose, endemic-rich Atlantic forest in Brazil (Chapter 1 frontispiece), and the red-cockaded woodpecker habitat in the pine forest of the southeastern USA (frontispiece of this chapter).

In Chapter 3, we addressed the harmful genetic effects on fitness, ability to evolve, and extinction risk of diminished habitat and population sizes. This chapter focuses on the genetic effects of population fragmentation per se, the separation of a population into partially or completely isolated fragments for the same total population size. Our primary focus is species whose habitats have been fragmented by human activities within the last 500 years (the period of greatest human impacts), rather than species that have naturally fragmented habitats.

The endangered red-cockaded woodpecker illustrates many of the genetic features associated with habitat fragmentation, including loss of genetic diversity within population fragments, and differentiation among populations (Box 4.1).

Box 4.1 Impact of habitat fragmentation on the endangered red-cockaded woodpecker in southeastern USA

(Frankham et al. 2017, Box 5.1)

The red-cockaded woodpecker was once common in the near-continuous mature pine forests of southeastern USA, but declined in numbers, primarily due to habitat loss, and now survives in isolated sites (black circles on map on chapter frontispiece) with ~ 1% of its original population size. The woodpeckers require old-growth forest, especially for nesting cavities.

As there is little gene flow among isolated sites, populations have lost genetic diversity and diverged genetically. Smaller, isolated populations show lower genetic diversity (and are more inbred) than larger ones, as expected (graph on chapter frontispiece). Further, more distant populations are on average more genetically diverged than nearby ones for molecular markers, and there are clines in wing and tail length associated with habitat temperature (Mengel & Jackson 1977). A recovery plan has been instituted to address genetic and other problems in the species, including enhancing gene flow between nearby populations, as described later in the chapter.

What do we mean by genetic fragmentation?

Our concern in this book is genetically distinct population segments within species that are partially to completely isolated genetically, and that have distinct evolutionary fates.

The human-induced isolating factors between population segments are typically inhospitable habitat, roads, rails, dams, agricultural lands, and livestock fencing (Coleman et al. 2018). Conservation fencing also isolates populations, as used to reduce human–wildlife conflicts with large cats and elephants in South Africa, and to protect threatened marsupials and rodents in Australia from introduced predators (Somers & Hayward 2012).

Adverse genetic effects of fragmentation arise from reduced gene flow

Fragmentation typically reduces gene flow among population fragments compared to continuous habitat, as for example in European beech trees (*Fagus sylvatica*: Fig. 4.1). Such fragmentation effects are widespread, but the scale over which they are evident depends upon distance between fragments, dispersal ability of species, and the suitability of the surrounding habitat matrix.

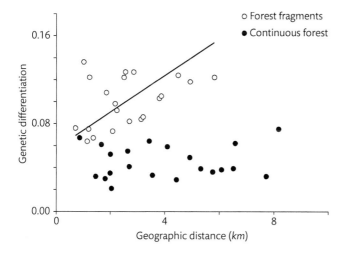

Fig. 4.1 Genetic differentiation among populations of European beech trees is greater and increases with distance in fragmented but not in continuous forest, because gene flow has been reduced by fragmentation (Frankham et al. 2017, Fig. 5.1, after Jump & Peñuelas 2006).

Dispersal typically declines with distance

Even in continuous habitats, dispersal rates in animals and plants decrease with distance (Fig. 4.2).

Since gene flow requires dispersal of individuals or gametes, genetic differentiation is often related to geographic distance, as observed in many animal and plant species.

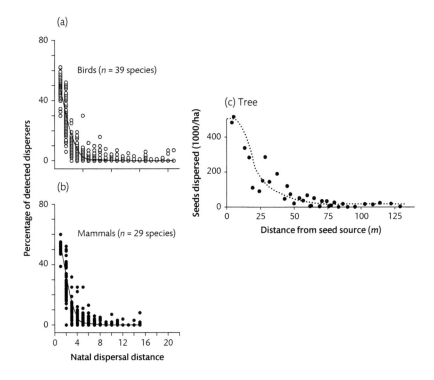

Fig. 4.2 Dispersal rates decline with distance in (a) birds, (b) mammals (both after Sutherland et al. 2000), and (c) a Eucalypt tree (from Cremer 1966, reprinted with permission from Taylor & Francis Ltd., from Cremer, K.W. (1966). Dissemination of seed from *Eucalyptus regnans*. *Australian Forestry*, 30(1), 33–37, http://www.informaworld.com) (Frankham et al. 2010, Fig. 14.11).

Genetic fragmentation depends upon dispersal ability and distance

Species with poor dispersal typically show greater differentiation among isolated population fragments than do species with good dispersal ability (Bohonak 1999). For example, vertebrates that walk show greater genetic differentiation with geographic distance than those that fly or swim (Fig. 4.3).

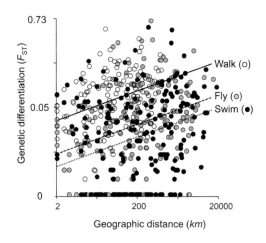

Fig. 4.3 Genetic differentiation among vertebrate populations that walk, fly, or swim plotted against standardized geographic distance, based on microsatellite data from 440 taxa (after Medina et al. 2018).

Many species exhibit isolation by distance (Fig. 4.3): for example, populations of bighorn sheep, gray wolves, and brown bears in North America, red-cockaded woodpeckers, northern spotted owls (*Strix occidentalis caurina*), and the thale cress plant (*Arabidopsis thaliana*).

Genetic problems due to inadequate gene flow are widespread

Approximately 29% of vertebrate, 26% of invertebrate, and 55% of plant species show clear evidence of genetically fragmented populations with inadequate gene flow, as determined from the proportion of species where populations exchange on average less than one contributing migrant per generation. Consequently, there are millions of populations with inadequate gene flow that are expected to be suffering genetic problems (Chapter 1).

Harmful genetic effects are worse in completely isolated fragments

The most extreme consequences of fragmentation occur when there is complete genetic isolation among population fragments. In this case, the effective population size determining inbreeding and loss of genetic diversity is that for each fragment, rather than for the species. Consequently, inbreeding and loss of genetic diversity occur more rapidly than in unfragmented populations and extinction risks are higher. Harmful genetic effects of population fragmentation under total isolation have been documented in several *Drosophila* fruit fly, house fly, and plant studies, and are almost ubiquitous in studies of fragmented outbreeding species in the wild. For example, a meta-analysis of plant studies showed that fragmented populations have less molecular genetic diversity than non-fragmented populations, and genetic diversity declines with generations of isolation (Fig. 4.4). Further, a second meta-analysis of animal and plant studies reached similar conclusions (Schlaepfer al. 2018).

Bighorn sheep

Gray wolf

Brown bear

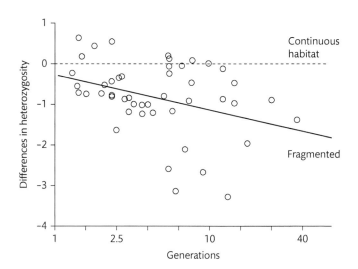

Fig. 4.4 Heterozygosity in fragmented plant populations declines with generations of isolation, based on data from 47 species (after Frankham et al. 2017, Fig. 5.9, based on Aguilar et al. 2008, Fig. 5). Standardized differences in heterozygosity between fragmented and continuous habitats (open circles) are plotted against generations.

We next discuss the effects of fragmentation on genotype frequencies, inbreeding, and genetic differentiation among fragments.

Population fragmentation reduces heterozygosity

As the allele frequencies in population fragments drift apart, the observed heterozygosity for the entire population declines when compared to Hardy–Weinberg expectations, referred to as the **Wahlund effect**. For example, this is observed in the Pacific yew (*Taxus brevifolia*) (Example 4.1 later in this chapter).

Heterozygosities lower than Hardy–Weinberg expectations are widely used to diagnose populations that are genetically fragmented.

Population fragmentation increases inbreeding

The increase in inbreeding coefficients in isolated population fragments over generations is faster than in a random mating population of the same total population size. For example, if a population with a size of $N_e = 100$ is split into five equal-sized, totally isolated population fragments of $N_e = 20$, inbreeding increases at a rate of 1/40 (2.5%) per generation in the small fragments versus 1/200 (0.5%) in the single large population. After 10 generations, the inbreeding coefficient is 40% in the small fragments but only 4.9% in the single large population (computed by the method in Example 3.1).

Isolated population fragments diverge genetically

When a population is fragmented by drift, different fragments will usually have slightly different initial allele frequencies. Diversification in allelic frequencies continues, generation after generation, until eventually the alleles in all populations are homozygous for different subsets of the original variation. This diversification is expected to:

- increase with generations (Fig. 4.5)
- increase faster in smaller than in larger population fragments (Fig. 4.6)
- reach a maximum when all populations have become homozygous.

Such variation in allele frequencies among population fragments has been found in virtually every species studied, one example being giant pandas as discussed in Box 1.1.

Similar increases in genetically determined morphological, behavioral, physiological, and disease resistance variation occur among population fragments due to genetic drift among animal and plant populations. For example, replicate isolated populations of guinea pigs (*Cavia porcellus*) founded from the same source population developed wide variation in coat color and patterns (Fig. 4.7), body weight, body length, internal organs, temperament, resistance to tuberculosis, and types and frequencies of morphological abnormalities (Wright 1977).

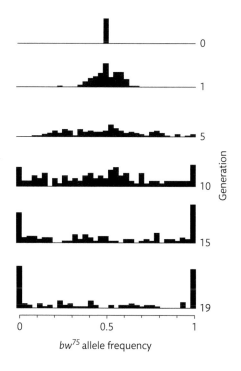

Generation

0

1

5

10

15

19

0   0.5   1

bw^{75} allele frequency

Fig. 4.5 Allele frequency divergence across generations for totally isolated population fragments of *Drosophila* fruit flies. The frequency distributions for the *bw*^{75} allele are shown across 19 generations in 105 replicate populations maintained with 16 parents per generation. All populations began with initial frequencies of 0.5 for the two alleles (Frankham et al. 2010, Fig. 14.6, after Buri 1956).

Genetic divergence (F_{ST})

1

N_e = 20

N_e = 50

N_e = 100

0.5

N_e = 500

0

0   100   200   300

Generation (*t*)

Fig. 4.6 Standardized genetic divergence among populations (F_{ST}) increases more rapidly over generations in populations with smaller effective population sizes (N_e), based on computer simulations (Frankham et al. 2010, Fig. 14.8).

Fig. 4.7 Variation in coat color and pattern among four isolated populations of guinea pigs, all derived from the same variable outbred base population and inbred for many generations by sib-mating (Wright 1977, pp. 70–71). The differences among populations were consistent and all animals within populations were relatively similar.

Population fragments simultaneously become genetically differentiated, lose genetic diversity, and become inbred due to drift

A critical, but little appreciated, expectation in conservation genetics is that population fragments simultaneously become genetically differentiated, lose genetic diversity, and become inbred due to drift: empirical observations with mice, marsupials, and fish have confirmed these expectations (Chapter 7).

Selection may alter the impact of fragmentation

Balancing selection typically reduces genetic differentiation among population fragments, while diversifying selection amplifies it

If alleles are subject to balancing selection, the rate of diversification will generally be lower than for neutral ones. Conversely, if an allele is favored in some fragments and harmful in others (diversifying selection), then the rate of diversification will be greater than predicted for neutral markers. Accordingly, genetic divergence among populations for quantitative characters is often elevated. However, if populations are small and selection is weak, beneficial and harmful alleles will often behave as if neutral.

Gene flow reduces the harmful genetic effects of population fragmentation

The genetic impacts of fragmentation (inbreeding, loss of genetic diversity, and divergence among populations) are inversely related to the rate of gene flow, also referred to as genetic connectivity. For example, resistance to a fungal disease in the plant *Plantago lanceolata* in Finland is higher for populations with greater gene flow.

But when is gene flow sufficient to prevent adverse genetic impacts of fragmentation, or to reverse them?

Low levels of gene flow per generation are sufficient to prevent adverse genetic impacts of fragmentation

About five or more successful immigrants per generation are required to avoid harmful inbreeding and loss of genetic diversity in fragmented populations (Chapter 8). The same number of contributing immigrants has similar effects in different sized populations. This may appear paradoxical, but one migrant that breeds represents proportionally a much higher rate of gene flow into a smaller than larger population, which counteracts the stronger genetic drift in smaller populations.

We specify successful immigrants because some may not breed, or be as successful in producing offspring as residents, so considerably more than five actual immigrants per generation is typically required. Further details are presented in Chapter 8.

The genetic consequences of different fragmented population structures vary

The genetic impacts of population fragmentation range from negligible to severe, depending on the population structure and gene flow among fragments. There are several models of fragmented population structure with gene flow (Fig. 4.8).

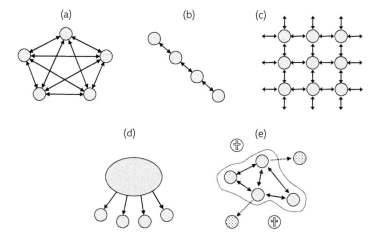

Fig. 4.8 Five different theoretical fragmented population structures (after Frankham et al. 2010, Fig. 14.1): (a) an island structure, where gene flow is equal among populations and unaffected by distance, (b) a linear stepping-stone structure, where only neighboring populations exchange migrants, (c) a two-dimensional stepping-stone structure, where neighboring populations in any direction exchange migrants, (d) a source–sink or mainland–island structure, where the sources provide all the input to the sinks (all after Hedrick 1983), and (e) a metapopulation with repeated extinctions (circles with crosses) and recolonizations (dotted circles) (after Hanski & Gilpin 1997).

These population structures differ in the harmfulness of their effects, and their relationships of genetic divergence with geographic distance. Isolation by distance is expected with stepping-stone population structures, but not with the island model where migration rates are equal between all populations. The effective size for a source–sink structure is that of the source only, while metapopulations may also have much smaller effective sizes than their total sizes, because genetic lineages are frequently lost through extinction of patches and populations are often recolonized by few individuals. For many natural populations, the metapopulation is a more realistic model than the others.

Metapopulations differ from the other fragmented structures in that there are frequent extinction and recolonization events (Fig. 4.9). For example, there are ~ 1,600 suitable meadows for Glanville fritillary butterflies on Åland Island in southwest Finland, 320–524 being occupied in 1993–1996, with an average of 200 extinctions and 114 colonizations per year.

The genetic consequences of a metapopulation structure are generally more harmful than the other population structures

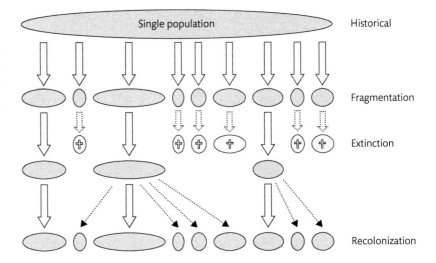

Fig. 4.9 Cycles of extinction and recolonization in a metapopulation lead to reductions in the effective population size of a species. The dotted lines indicate bottlenecks during recolonizations and the circles with crosses extinctions (Frankham et al. 2010, Fig. 14.14).

In general, genetic diversity within populations decreases with higher rates of extinction and recolonization in a metapopulation, and inbreeding increases as the colonists to a particular location become more related to one another.

We next address means for measuring genetic differentiation among populations.

How do we measure genetic differentiation among populations?

F statistics

We begin by describing *F* statistics, a commonly used measure of population differentiation with many derivatives. Inbreeding in the total (*T*) population (F_{IT}) can be partitioned into inbreeding of individuals within their population fragment F_{IS}, and inbreeding due to differentiation among population fragments F_{ST}. These measures of inbreeding reflect departures from random mating (Hardy–Weinberg equilibrium) within and among populations, as well as divergence in allele frequencies among population fragments. *F* statistics are usually estimated from heterozygosity of genetic markers, as illustrated in Example 4.1 for the rare Pacific yew tree in Canada.

F_{ST}, the inbreeding due to differentiation among populations, is our primary focus here. It is estimated as the proportionate deviation of Hardy–Weinberg heterozygosities over fragments compared to the Hardy–Weinberg heterozygosity for the total population. F_{ST} was originally defined in terms of standardized variance in allele frequencies, which better reflects its use to measure genetic differentiation, but the two definitions are equivalent when dealing with drift alone. Unfortunately, F_{ST} does not scale 0–1 when there are more than two alleles for a gene, so values obtained from using different markers are often not comparable.

F_{IS} is the proportional deviation of the observed heterozygosity within fragments from the Hardy–Weinberg expected heterozygosity of each fragment. This is not recommended as a general measure of inbreeding, as it only reflects deviations from random mating in the most recent generation, rather than total inbreeding over generations.

Example 4.1 F statistics for populations of the Pacific yew
(Frankham et al. 2010, Example 14.4, based on El-Kassaby & Yanchuk 1994)

Average observed heterozygosity (H_o) for 21 allozyme genes across nine Canadian populations was 0.085, while the average expected heterozygosity for these populations (H_S) was 0.166, and the expected heterozygosity for the nine populations combined (H_T) was 0.18.

Inbreeding due to population differentiation is:

$$F_{ST} = 1 - \frac{H_S}{H_T} = 1 - \frac{0.166}{0.180} = 0.078$$

This indicates only a modest degree of population differentiation, as F_{ST} ranges from 0 to 1 when there are only two alleles for each gene.

Inbreeding within populations is:

$$F_{IS} = 1 - \frac{H_o}{H_e} = 1 - \frac{0.085}{0.166} = 0.49$$

This inbreeding is not due to selfing because the species has separate sexes, but is probably due to offspring establishing close to parents in bird and rodent seed caches, and subsequently more mating between close relatives than expected by chance.

The total inbreeding across the population fragments due to the combination of the two effects is:

$$F_{IT} = 1 - \frac{H_o}{H_T} = 1 - \frac{0.085}{0.180} = 0.53$$

Pacific yew (North America)

There are a number of problems associated with using F_{ST} for conservation purposes that we discuss in Chapter 8. Therefore, we now introduce the concept of kinship, as it is a more useful parameter for the genetic management of fragmented populations than F_{ST}, scales 0–1, and provides an alternative measure of population differentiation without the problems associated with F_{ST}.

Kinship

Kinship (k_{ij}, also called coancestry) is based on how related pairs of individuals are, and it is the inbreeding coefficient of an offspring if they had one. Thus, it always scales 0–1, regardless of the number of alleles per marker. Mean kinship within a population is simply the mean of all the pairwise kinship values between individuals, including kinships with self. Mean kinship between two populations A and B (mk_{AB}) can be used to measure population similarity, so ($1 - mk_{AB}$) provides a measure of population genetic differentiation. Mean kinship is widely used for genetic management of captive populations of threatened species, and in Chapter 8 we show that it is superior to F_{ST} for genetic management of wild populations as well. Kinship can be measured from pedigrees or estimated from data on multiple genetic markers such as microsatellites or SNPs.

The number of populations can be estimated from the number of genetic clusters

If groups of individuals occur further apart than the distance over which mating and dispersal occur, then they are likely to belong to different genetic populations. However, such an approach is not definitive because high levels of gene flow may result in physically isolated population fragments acting effectively as a single genetic population. Further, geographic isolation may vary over time, as for example with boom and bust cycles of small mammals inhabiting arid areas, or freshwater fish and other aquatic species affected by cycles of droughts and floods that result in once isolated populations becoming connected and vice versa.

As individuals may exist in a continuum with approximately random mating, or show isolation by distance, clines, or multiple genetically isolated populations, we need to distinguish the patterns in Fig. 4.10. In practice, such patterns are often modified by landscape features that may inhibit or facilitate gene flow such as rivers, lakes, mountains, soil changes, roads, and fences. Another common pattern is isolation by environment, where gene flow is greater among more similar environments.

Fig. 4.10 Schematic view of population genetic structures we need to distinguish: (a) single random mating population, (b) isolation by distance, (c) isolation by distance broken by recent habitat fragmentation, (d) cline where gene flow reduces over an environmental gradient (e.g. rainfall), often associated with adaptive differences, (e) cline broken by recent habitat fragmentation, and (f) geographically and genetically isolated populations (Frankham et al. 2017, Fig. 10.2). Individuals within the boundaries are connected by gene flow. The different intensities of shading indicate genetic differentiation along geographic or environmental gradients.

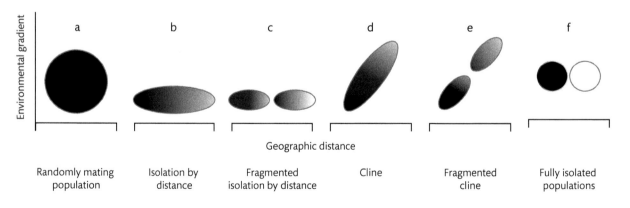

The question of whether or not there are multiple genetic populations can be answered by genotyping individuals and asking if there are genetic clusters (or another of the above pattern), as illustrated for the Mexican wolf and related canids in Box 4.2.

Box 4.2 Identifying the number of distinct populations of Mexican wolf, whether there has been gene flow from other taxa, subsequent management actions, and their outcomes

(Frankham et al. 2017, Box 10.1, based on Hedrick et al. 1997; Fredrickson et al. 2007; Hedrick & Fredrickson 2010)

Three small inbred populations of presumed Mexican wolves (a sub-species of the gray wolf) were all that remained in the 1990s. Could they be combined to alleviate problems of low genetic diversity, and inbreeding depression? Managers needed to know whether the three populations were "pure" Mexican wolves, or whether they had crossed with other canid taxa (introgression). A cluster analysis approach (multidimensional scaling) based on allele frequencies for 20 microsatellites showed that the three Mexican wolf populations formed a cluster distinct from other canids. Thus, all three populations of Mexican wolves were "pure," rather than showing evidence of introgression from other canids (graph below).

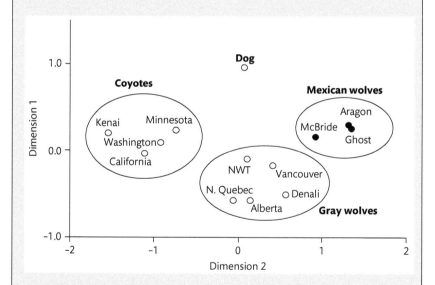

Mexican wolf

(from Frankham et al. 2017, p. 206, after Hedrick et al. 1997)

Consequently, the three populations were combined, resulting in substantial improvements in fitness and genetic diversity.

The most widely used software packages for determining whether there are single or multiple populations are based on random mating within but not among population segments

STRUCTURE, a commonly used software package, implements a clustering algorithm on genetic marker data (typically microsatellites or SNPs) from population samples of individuals to estimate the number of random mating genetic populations. It then assigns each individual to one or more of those genetic populations, as shown for giant panda populations in Box 1.1.

Isolated populations with low genetic diversity and strong genetic divergence from other populations are relatively easily distinguished by STRUCTURE and related methods, given adequate marker data and sampling. Simulations showed that STRUCTURE yielded the correct clusters down to $F_{ST} = 0.03$. For perspective, European human populations have $F_{ST} = 0.02$, European and Chinese have $F_{ST} = 0.11$, and Europeans and Africans $F_{ST} = 0.15$.

However, to manage populations, we must locate them in the environment, and STRUCTURE does not do this for us (unless we know the location where each individual was collected).

Locating genetically differentiated populations in the environment

Analyses of patterns of genetic diversity across the environment can locate differentiated populations geographically, delineate clines and boundaries to gene flow, and detect isolation by distance, metapopulations, and randomness

Geographic fragmentation within species can be defined by mapping patterns of genetic diversity onto geography. Software such as GENELAND and TESS achieve this using genotypic clustering on georeferenced genetic data from population samples of individuals. For example, GENELAND was used to identify and locate genetically differentiated populations of the Florida scrub-jay (*Aphelocoma coerulescens*) and relate their presence to habitat and geographic features (Fig. 4.11). Genomic analyses showed that a reduction in immigration into the population fragments from 1995 to 2013 resulted in increased levels of inbreeding and reduced fitness via inbreeding depression, even though the population sizes remained similar (Chen et al. 2016).

Where possible, we recommend the use of georeferenced data and analyses.

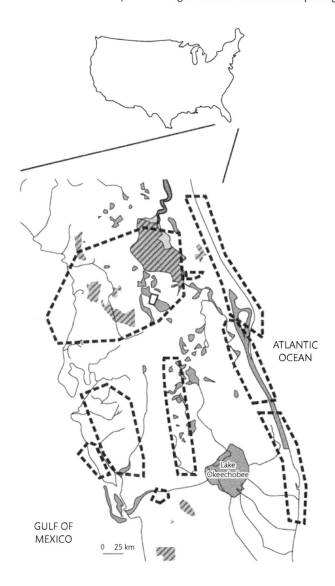

ATLANTIC
OCEAN

Lake
Okeechobee

GULF OF
MEXICO

0   25 km

Fig. 4.11 Genetic distinctiveness mapped onto geographical regions using GENELAND for the threatened Florida scrub-jay, based on genotypes of > 1,000 birds for 20 microsatellites. The heavy dotted lines represent genetically differentiated populations, crosshatched areas are forest, and solid gray areas are lakes and rivers. This genetic structuring of the populations corresponds roughly to where habitat features (such as rivers and lakes) and unsuitable habitat isolate small populations and where large extents of suitable habitat allow more connectivity. Coastal populations are genetically differentiated from each other and from the interior populations, despite the short geographical distances separating them (Frankham et al. 2017, Fig. 10.4, after Coulon et al. 2008, Fig. 5).

Florida scrub-jay (USA)

In the next chapter we consider genetic rescue of population fragments suffering genetic erosion by gene flow from one or more other populations. Further, we consider the potential risk that such crossing will be harmful, rather than beneficial.

Summary

1. Most species have fragmented distributions, often with limited gene flow between population fragments.
2. Population fragmentation with cessation of gene flow leads to greater inbreeding, lower fitness, more rapid loss of genetic diversity, less evolutionary adaptation, and elevated extinction risk, as compared to a single unfragmented population of the same total size.
3. Genetically diverged population fragments have reduced genetic diversity within them, and increased inbreeding.
4. The rates at which fragmented populations diverge in allele frequencies, lose heterozygosity, and become inbred increase as effective population sizes and levels of gene flow decrease, and the adverse impacts increase with generations.
5. Low levels of gene flow per generation are generally sufficient to prevent the harmful genetic effects of fragmentation.
6. The number and location of population fragments can be estimated by clustering individuals based on georeferenced data on multiple genetic markers.

FURTHER READING

Aguilar et al. (2008) Meta-analysis on the genetic impacts of habitat fragmentation in plants.

Frankham et al. (2017) *Genetic Management of Fragmented Animal and Plant Populations*: Chapters 5 and 10 have more detailed treatments of topics in this chapter.

Phillipsen et al. (2015) Theoretical prediction and empirical data on genetic divergence with geographic distance for different gene flow rates.

SOFTWARE

GenAlEx: computes basic population genetic statistics, including F statistics, and tests for isolation by distance (Peakall & Smouse 2006). http://biology-assets.anu.edu.au/GenAlEx/Download.html

GENELAND: a program for identifying population units and delineating their landscape locations, akin to a spatial version of STRUCTURE (Guillot et al. 2005). https://i-pri.org/special/Biostatistics/Software/Geneland/

GENEPOP: computes F statistics, and tests for isolation by distance (Rousset 2008). http://kimura.univ-montp2.fr/~rousset/Genepop.htm

PMx: computes kinships, mean kinships, and between population mean kinships from pedigrees (Ballou et al. 2018). https://scti.tools/pmx

STRUCTURE 3.2: frequently used software to delineate number of populations, based on Bayesian clustering (Falush et al. 2007). https://web.stanford.edu/group/pritchardlab/structure_software/release_versions/v2.3.2/html/structure.html

Genetic rescue resulting from gene flow

CHAPTER 5

Inbreeding is reduced and genetic diversity enhanced when a small isolated inbred population is crossed to an unrelated population. Crossing can have beneficial or harmful effects on fitness, but benefits predominate, and the risks of harm (outbreeding depression) can be predicted and avoided. For crosses with a low risk of outbreeding depression in species that naturally outbreed, large and consistent fitness benefits persist across generations. Benefits are greater in species that naturally outbreed than in those that inbreed, increase with greater disparity in inbreeding level between crossed and inbred populations, and are higher when immigrants are outbred than when they are inbred. Benefits are similar across invertebrates, vertebrates, and plants. Gene flow into populations with low genetic diversity also improves their ability to evolve.

TERMS

Centric fusion, chromosomal translocation, coadapted gene complexes, evolutionary rescue, heterosis, inversion, polyploid, sub-species, tetraploid

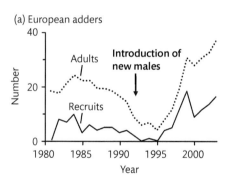

(a) European adders

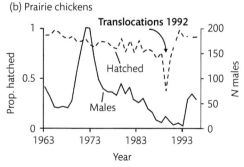

(b) Prairie chickens

Declines in numbers and fitness due to inbreeding and recovery through genetic rescue following subsequent introduction of immigrants (arrowed) in (a) European adders and (b) greater prairie chickens (North America) (Frankham et al. 2017, p. 115, after Madsen et al. 2004; and Westemeier et al. 1998, respectively).

A Practical Guide for Genetic Management of Fragmented Animal and Plant Populations. R. Frankham, J. D. Ballou, K. Ralls, M. D. B. Eldridge, M. R. Dudash, C. B. Fenster, R. C. Lacy & P. Sunnucks. Oxford University Press (2019). © R. Frankham, J. D. Ballou, K. Ralls, M. D. B. Eldridge, M. R. Dudash, C. B. Fenster, R. C. Lacy & P. Sunnucks 2019. DOI: 10.1093/oso/9780198783411.001.0001

Genetic erosion occurs in many small isolated population fragments

Small isolated population fragments lose genetic diversity and become inbred over generations, leading to inbreeding depression, reduced ability to evolve, and increased extinction risk (Chapters 2–4). In this chapter, we show that these adverse genetic effects can often be reversed by gene flow (also referred to as genetic rescue or crossing). For example, Box 5.1 describes genetic rescue of the inbred Florida panther population following gene flow from the Texas population.

Box 5.1 Genetic rescue of inbred Florida panthers following gene flow

(after Frankham et al. 2017, Box 6.1, based on Roelke et al. 1993; Driscoll et al. 2002; Hostetler et al. 2010, 2012; Johnson et al. 2010; Benson et al. 2011)

Prior to European settlement, the Florida panther ranged across the entire southeastern USA, and other sub-species were spread throughout North and South America. By the early 1990s, Florida panthers had declined to a small relict population of ~ 20–25 individuals in southern Florida. Two groups of panthers existed in Florida, one unhybridized remnant and another hybridized with released animals from a South America sub-species.

The unhybridized Florida panthers had very low levels of genetic diversity for several genetic markers compared to earlier museum specimens, other populations of the species, and other felid species. They were highly inbred, and displayed evidence of inbreeding depression, including kinked tails, cardiac defects, a high prevalence of infectious disease, low numbers of motile sperm, lowered testosterone levels, and a high incidence of males with at least one undescended testis (cryptorchidism).

In 1995, the Florida population was augmented with eight wild-caught Texan females, whose population was historically contiguous and genetically connected. Five of the eight introduced Texas pumas bred and produced offspring. The resulting admixed population had ~ 50% higher heterozygosity than the original population, had tripled in numbers by 2007, mostly due to the success of hybrid individuals, and reached 230 panthers by 2017 (Flescher 2017). Hybrid offspring had higher adult, sub-adult, and cub survival, and higher litter size, resulting in a 169% higher total fitness than in the inbred parent population. Admixed offspring also had reduced levels of kinked tails, cardiac defects, and cryptorchidism, as well as improved sperm quality.

Florida panther (USA)

Inbreeding and loss of genetic diversity can be reversed by gene flow from another population

If two completely inbred populations ($F = 1$) homozygous for different alleles at a genetic marker are crossed, F_1 individuals have an inbreeding coefficient of zero and all are identically heterozygous (Table 5.1). However, in the F_2 individuals the inbreeding coefficient increases to ½ and the heterozygosity halves. If only some of the parental inbred individuals are crossed, the inbreeding coefficient will be reduced by a correspondingly smaller proportion in generation 1 and increase at a faster rate over subsequent generations than when all parents are crossed. Benefits of crossing are greater if an inbred population is crossed to outbred rather than other distinct inbred individuals (see later in this chapter).

Table 5.1 Reduced inbreeding and increased genetic diversity following crossing of distinct completely inbred populations. Inbreeding coefficients (F) and heterozygosity across generations are shown (Frankham et al. 2017, Table 6.1).

Population	Genotypes	F	Heterozygosity
Inbred a	A_1A_1	1	0
Inbred b	A_2A_2	1	0
F_1 (a × b)	A_1A_2	0	1
F_2 (F_1 × F_1)	¼ A_1A_1: ½ A_1A_2: ¼ A_2A_2	½[a]	½[a]

[a] These will remain the same in subsequent generations with random mating in large populations.

But what effect does crossing have on fitness?

Crossing can have beneficial or harmful effects on fitness

Many managers have been reluctant to undertake genetic rescue due to fear that crossing would result in outbreeding depression, rather than genetic rescue. However, we have devised an effective procedure to determine the risk of outbreeding depression, based on its known causes (Frankham et al. 2011). This allows the benefits of genetic rescue to be exploited. There is a low risk of outbreeding depression for crosses of populations with no fixed chromosomal differences, that are adapted to similar environments, and were previously connected by gene flow.

We next show that the benefits of genetic rescues on fitness and ability to evolve are large and consistent enough to have practical use in biodiversity conservation. We defer details of outbreeding depression until later in this chapter.

Fitness benefits from genetic rescues are large and consistent

This section focuses on circumstances relevant to practical conservation attempts to genetically rescue small inbred populations, namely the impacts of crossing an inbred population to another population when the risk of outbreeding depression is low. We present the impacts of crossing on fitness as the relative difference of fitness for the crossed progeny compared to the inbred parent (referred to as genetic rescue ΔGR: Box 5.2).

Box 5.2 Measuring the impact on fitness of interpopulation crossing
(Frankham et al. 2017, Box 6.2)

We measure the impacts on fitness of crossing inbred populations as the relative difference caused by genetic rescue (ΔGR), as follows:

$$\Delta GR = \frac{(\textit{fitness of crossed population} - \textit{fitness of inbred population})}{\textit{fitness of inbred population}}$$

Thus, a value > 0 represents beneficial effects of crossing, 0 means equal fitness of the crossed and inbred parent generations, and negative values indicate outbreeding depression.

For example, inbred African lions in Hluhluwe-iMfolozi Park in South Africa produced an average of 0.465 weaned cubs per female, but this increased to 2.077 following gene flow from Namibian lions (Trinkel et al. 2008). Thus, ΔGR is (2.077 – 0.465)/0.465 = 3.47, a 347% improvement.

Conversely, in the largely selfing poorjoe plant (*Diodia teres*), pollination success was 47.6% for the inbred parent populations, but 44.7% for F_1 population crosses (Hereford 2009), yielding ΔGR of (44.7 – 47.6)/47.6 = – 0.063, a 6.3% decline.

Inbreeding depression is usually reversed by crossing between populations (genetic rescue)

Crossing of populations has frequently been used to successfully genetically rescue laboratory and agricultural species, where rescue is referred to as heterosis, or hybrid vigor. For example, heterosis has been fundamental to maize production in the USA, and the "green revolution" that has substantially increased yields from several other crop plants (Evenson & Gollin 2003). Similarly, breed or strain crosses are often used to improve productivity of chickens (*Galus gallus domesticus*) and pigs (*Sus scrofa*). More recently, gene flow has been used to recover fitness in small inbred populations of several species of wild animals and plants (some of conservation concern), including greater prairie chickens and European adders (chapter frontispiece), bighorn sheep, deer mice (*Peromyscus maniculatus*), Florida panthers (Box 5.1), gray wolves, African lions, Mexican wolves, mountain pygmy possums (*Burramys parvus*), desert topminnow fish, Florida ziziphus, jellyfish trees, and partridge pea plants (*Chamaecrista fasciculata*) (Weeks et al. 2017).

Gene flow results in large and consistent fitness benefits

One impediment to genetic rescue attempts has been the lack of a quantitative overview of its impacts on fitness. A recent meta-analysis revealed that gene flow into inbred populations was beneficial in 92.9% of 156 cases across invertebrates, vertebrates, and plants. Benefits were reported for a variety of fitness measures, including total fitness, female and male lifetime reproductive success, female fecundity, survival, population growth rate, final population size, gamete quality, and fertilization success.

The benefits are even greater for natural outbreeding than inbreeding species, 162% in stressful and 51% in benign environments. Selected examples of genetic rescue effects on fitness are presented in Table 5.2. For example, there was a 151% genetic rescue effect for total fitness when a small, inbred population of the endangered jellyfish tree was crossed to another population in the Seychelles.

> Median benefits from deliberate gene flow were 148% in stressful and 45% in benign environments

Table 5.2 Magnitude of genetic rescue effects (ΔGR) for reproductive fitness following gene flow into small, isolated, inbred populations for a sample of species. ΔF is the reduction in inbreeding due to gene flow (after Frankham 2015).

Taxon	ΔGR (%)	Trait	ΔF	Mating system[a]
Vertebrates				
African lion (*Panthera leo*)	347	Cubs weaned/female	–[b]	O
Bighorn sheep (*Ovis canadensis*)	331	Female annual reproductive success	0.25	O
European tree frog (*Hyla arborea*)	15	Tadpole body mass	0.15	O
Greater prairie chicken (*Tympanuchus cupido pinnatus*)	26	Hatching success	0.10	O
Gray wolf (*Canis lupus*)	23	Annual population growth	0.14	O
Mexican wolf (*Canis lupus baileyi*)	184	Composite fitness	≥ 0.17	O
Song sparrow (*Melospiza melodia*)	47	Mean of female and male lifetime reproductive success	0.07	O
South Island robin (*Petroica australis*)	679	Reproductive recruitment per egg	0.21	O
European adder (*Vipera berus*)	233	Male recruitment success	0.75	O
Invertebrates				
Drosophila fruit fly (*Drosophila melanogaster*)	114	Total fitness	0.19	O
Glanville fritillary butterfly (*Melitaea cinxia*)	211	Egg hatching rate	0.41	O
Mysid shrimp (*Americamysis bahia*)	318	Net increase in N	0.31	O

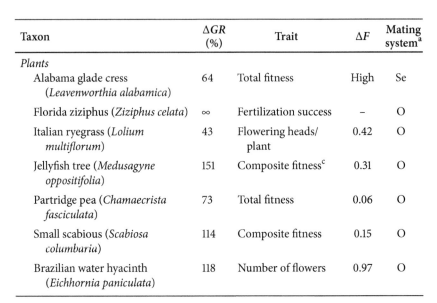

Taxon	ΔGR (%)	Trait	ΔF	Mating system[a]
Plants				
Alabama glade cress (*Leavenworthia alabamica*)	64	Total fitness	High	Se
Florida ziziphus (*Ziziphus celata*)	∞	Fertilization success	–	O
Italian ryegrass (*Lolium multiflorum*)	43	Flowering heads/ plant	0.42	O
Jellyfish tree (*Medusagyne oppositifolia*)	151	Composite fitness[c]	0.31	O
Partridge pea (*Chamaecrista fasciculata*)	73	Total fitness	0.06	O
Small scabious (*Scabiosa columbaria*)	114	Composite fitness	0.15	O
Brazilian water hyacinth (*Eichhornia paniculata*)	118	Number of flowers	0.97	O

[a] O = natural outbreeding species; Se = selfing.
[b] Not available (unknown).
[c] A fitness measure with contributions from survival and reproduction data, but with less information than contributing to total fitness.

Jellyfish tree (Seychelles)

Changes in fitness from augmenting gene flow varied from –14% to infinity, the latter being a population of Florida ziziphus, a self-incompatible plant species (incapable of self-fertilizing) with populations incapable of producing seed until they were crossed to other populations.

Variables affecting the magnitude of genetic rescue for fitness

As genetic rescue (in the absence of outbreeding depression) involves reversal of inbreeding depression, the variables affecting it are those that affect inbreeding depression, but with effects in the opposite direction. Thus, we observe that genetic rescue is (Table 5.3):

- increased as the disparity between inbreeding levels in the parental and crossed generations increases (ΔF)
- larger in wild or stressful environments than in benign ones
- greater when immigrants are outbred, rather than inbred
- greater in naturally outbreeding species than in ones with elevated frequencies of self-fertilization
- observed in vertebrates, invertebrates, and plants.

The combined effects of several of these variables on genetic rescue are illustrated by the classic study of a small inbred population of desert topminnow fish. A 75-fold increase in total fitness was observed, one of the largest benefits recorded. It involved several conditions expected to yield benefits, namely a naturally outbreeding species, a

Desert topminnow fish (Mexico)

Table 5.3 Tests for effects of different variables on the magnitude of genetic rescue (ΔGR) (Frankham 2015).

Variable	Median ΔGR (%)	n^{a}
Mating system	Outbreeding > selfing[***]	133
Outbreeding	78.8	
Selfing or mixed mating	16.5	
Environment (outbreeders)	Stressful > benign[***]	114
Stressful/wild	113.9	
Benign	48.0	
Immigrants	Outbred > inbred[***]	120
Outbred	113.6	
Inbred	51.9	
Major taxa	n.s.	133
Invertebrates	58.4	
Vertebrates	94.2	
Plants	59.1	

[a] Number of comparisons.
[***] $P < 0.001$.
n.s. Non-significant.

parental population that was highly inbred and exhibiting substantial inbreeding depression, measurement of total fitness in stressful wild conditions, and outbred immigrants.

Genetic rescue effects typically persist across generations in outbreeding species

Fitness effects of crossing in outbreeding species are expected to stabilize by the F_3 generation, because reduction in inbreeding is stable beyond this point under large effective population sizes in outbreeding species. Empirical results show that the fitness benefits of genetic rescue persist across generations: the percentages of beneficial cases were similar in the F_1, F_2, and F_3 and later generations based on a meta-analysis (Table 5.4). Median fitness effects of gene flow were significantly beneficial in the F_1, F_2, and F_3 generations, and F_2 and F_3 benefits were at least as great as those in the F_1.

Table 5.4 Percentage of beneficial comparisons and median benefits in offspring resulting from crossing inbred populations in the F_1, F_2, and F_3 (and later) generations (Frankham 2016).

Generation	Beneficial (%)	Median benefits (%)
F_1	90.5[**]	42.0[**]
F_2	100.0[**]	84.0[*]
F_3 and later	94.1[**]	86.0[**]

* $P < 0.05$; ** $P < 0.01$ for benefit compared to no effect.

Fitness benefits in outbreeding species persist beyond the F_3 generation, as there was no decline between generations F_3 and F_{16}, based on the data set for the study described in Table 5.4. However, if the population size of the crossed population remains small, it will become inbred again, and once again lose fitness.

Conversely, in species that habitually self-fertilize, heterozygosity in the crossed populations halves in each succeeding generation after the F_1 until there is none, inbreeding levels increase correspondingly, and thus fitness benefits are expected to decay.

Inbreeding depression can be prevented with repeated immigration over generations

Regular low levels of immigration can prevent fitness declines in small isolated populations. This is expected to greatly reduce inbreeding and has been shown to minimize inbreeding depression in plants and animals. For example, inbreeding rose rapidly over six generations in isolated populations of field mustard (*Brassica campestris*) maintained with population sizes of five individuals, but 2.5 immigrants per generation prevented inbreeding from increasing, whereas populations with a single immigrant per generation had an F of 0.08 at generation 6 (Fig. 5.1a). Both immigrant treatments had much higher fitness than the totally isolated populations (Fig. 5.1b). Similarly, small bottlenecked house fly populations maintained with either 0, 1, or 20 immigrants per generation for 24 generations had larval survivals of 22%, 43%, and 53%, respectively, in the final generations. Further, 66% of the totally isolated populations went extinct, whereas all populations with regular immigration survived.

Fig. 5.1 Effects of immigration in field mustard. (a) Observed (pedigree) and expected inbreeding coefficients (*F*) over generations for the 0, 1, and 2.5 immigrants per generation treatments and (b) seed numbers for plants of the three immigration treatments at generation 6 in an outdoor common garden (Frankham et al. 2017, Fig. 6.1, after Newman & Tallmon 2001).

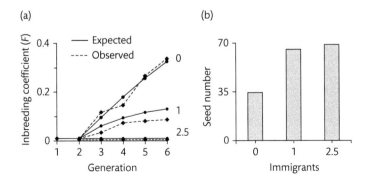

So far, we have focused primarily on the fitness benefits of crossing. However, the ability to evolve is also expected to rise with increasing genetic variation following gene flow.

Gene flow enhances the ability to evolve (evolutionary rescue)

Empirical studies showed that 65 of 72 of evolutionary rescue attempts were beneficial, and the few exceptions had low statistical power (Frankham 1980). For example, crosses of replicate *Drosophila* fruit fly populations previously maintained with an effective size of 50 for 50 generations had 79% higher ability to evolve than their inbred parent populations when all were subjected to increasing concentrations of NaCl. This is crucial in the context of environmental change (Pelletier & Coltman 2018; Chapter 9).

Box 5.3 illustrates the potential use of gene flow to minimize the devastating impacts of toxic cane toads (*Rhinella marina*) on the northern quoll (*Dasyurus hallucatus*), a small Australian endemic carnivorous marsupial.

Box 5.3 Potential use of evolutionary rescue to minimize the impacts of invasive cane toads on the northern quoll
(Kelly & Phillips 2016, 2019a, 2019b)

A relatively new use for evolutionary rescue is to introduce genetic diversity for traits that would benefit populations threatened by spreading invasive species. This has been proposed as a means for mitigating the impacts of invasive toxic cane toads on northern quolls in northern Australia. As cane toads spread they devastate quoll populations because the naive quolls consume toads and die. However, some quoll populations have evolved to avoid eating toads. It has been proposed that these "toad smart" quolls be introduced into naïve quoll populations ahead of the cane toad invasion front to provide genetic material to enable hybrid quolls to persist when the toads arrive, rather than the local quolls being extinguished.

Modeling suggests that introduction of 25 or more "toad smart" quolls in the year (~ 1 generation) prior to the arrival of the toads is expected to prevent quoll extinctions, as there will be "toad smart" hybrid animals present when the toads arrive.

Such an approach has promise for other species impacted by toxic invasive animals (e.g. snakes, lizards, and birds), provided the species have populations that have evolved tolerance to the toxin or avoidance of its consumption.

Northern quoll (Australia)

Evolutionary rescue depends on increase in genetic diversity and intensity of selection

The ability of an inbred population to adapt evolutionarily is enhanced as more genetic diversity is added (Fig. 5.2): all these data came from invertebrates, but similar results are expected for vertebrates and plants.

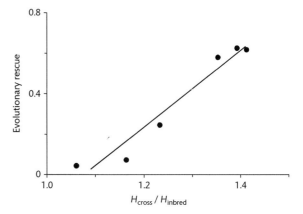

Fig. 5.2 Evolutionary rescue for fitness traits of formerly inbred populations increases with the degree of enhancement of genetic diversity (represented here as H_{cross} / H_{inbred}). Evolutionary rescue is presented here as the ratio of evolutionary change per generation in the cross to that in the inbred, plotted on a $\log_e$ scale. The line of best fit is shown (Frankham et al. 2017, Fig. 6.2, after Frankham 2015, Fig. 2).

Because crossing increases fitness and therefore the intensity of selection (i.e. there will be a larger pool of offspring produced on which selection can act each generation), there will be additional evolutionary rescue benefits beyond those due to the increase in heterozygosity.

We now identify the primary mechanisms causing outbreeding depression as a prelude to explaining how we can predict its risk.

The mechanisms generating outbreeding depression are known

Three primary mechanisms are known to generate outbreeding depression in within species crosses:

1. Fixed chromosomal differences
2. Adaptive differences among populations
3. Coadapted gene complexes (ones where certain combinations of alleles from different genes function best)

Table 5.5 lists examples of crosses that exhibit or do not exhibit outbreeding depression, along with information on characteristics that are expected to predict the risk of outbreeding depression. In what follows, we concentrate primarily on the causes of outbreeding depression within species.

Table 5.5 Examples of species and population crosses that have and have not exhibited outbreeding depression, and associated characteristics likely to predict adverse consequences (after Frankham et al. 2017, Table 7.1).

Taxa	Observed outbreeding depression	Taxonomic status of populations	Fixed chromosome differences	Gene flow in last 500 years	Environments
Corroboree frog (*Pseudophryne corroboree* and *P. pengilleyi*)	Yes	2 species	Unlikely	No	Similar
Ibex (*Capra ibex ibex*)	Yes	3 species[a]	Unlikely	No	Different
Owl monkey (*Aotus trivirgatus*)	Yes	Some consider them 2 species	Yes	Unlikely	Unknown
Kirk's dik dik (*Madoqua kirki*)	Yes	>1 species	Yes	Unlikely	Unknown
Button wrinklewort (*Rutidosis leptorrhynchoides*)	Yes	Called 1 species	Yes, hexaploid, tetraploid, and diploid	Unlikely	Similar
Peromyscus (*P. polionotus leucocephalus* × *P. p. subgriseus* or *P. p. rhoadsi*)	Yes, modest	Beach mouse sub-species × old-field mouse sub-species	No	No	Different
Peromyscus (*P. p. subgriseus* × *P. p. rhoadsi*)	No	Sub-species of old-field mouse	No	Likely	Similar
Partridge pea (*Chamaecrista fasciculata*)	Yes	Crosses between distant populations	No	Unlikely	Different
Copepod (*Tigriopus californicus*)	Yes	Likely a species complex	No	No	Some different, some similar
Florida panther (*Puma concolor*)	No	1 species[b]	No	Yes	Moderately different
Pink salmon (*Oncorhynchus gorbuscha*)	Unclear	1 species	No (odd and even years have 52 vs 52, 53, and 54 chromosomes)	No	Similar[c]
Mountain pygmy possum[d] (*Burramys parvus*)	No	1 species	No	No	Similar
Golden lion tamarin (*Leontopithecus rosalia*)	No	1 species	No	Yes	Similar

[a.] Populations used are uncertain, but probable candidates are now classified as separate species.
[b.] Previously classified as two sub-species, but subsequently revised based on molecular studies.
[c.] Distributions partially different so may be adapted to partially different environments.
[d.] Weeks et al. (2017).

We now discuss the three primary mechanisms causing outbreeding depression. Note that we use the terms outbreeding depression and reproductive isolation interchangeably to describe reductions in reproductive fitness in the progeny of population crosses in the F_1 or later generations.

Mechanism 1: Fixed chromosomal differences

Fixed chromosomal differences between populations are well-established causes of reproductive problems when such populations are crossed. For example, horses (*Equus caballus*) and donkeys (*Equus asinus*) have 31 and 32 pairs of chromosomes that differ by many translocations, inversions, etc., and their cross results in sterile mules. The major types of chromosomal differences involved in causing outbreeding depression are:

- ploidy mismatches
- structural changes in chromosomes (translocations, inversions, and centric fusions).

When populations with such structural chromosomal differences are crossed, the F_1 progeny produce some gametes with portions of the genome missing and other parts duplicated (unbalanced chromosomal constitutions) and zygotes resulting from these gametes typically do not survive or are abnormal.

The severity of outbreeding depression is usually greatest with ploidy mismatches, next most severe in the presence of chromosomal translocations or complex centric fusions, followed by simple centric fusions and inversions, as detailed in Frankham et al. (2017). Adverse effects typically increase with the number of fixed chromosomal differences.

Ploidy differences

Many plant species have populations with different ploidy levels, yet these are still referred to as the same species, despite crosses between them being harmful. For example, the endangered button wrinklewort daisy in eastern Australia has diploid, tetraploid, and hexaploid populations with similar phenotypes (Fig. 5.3). Progeny of crosses between diploid and tetraploid forms are triploid and usually sterile because meiosis produces gametes with a great diversity of chromosome numbers. In general, crosses between different ploidy levels result in substantial sterility.

Button wrinklewort daisy (Australia)

Polyploid populations have chromosome numbers that are multiples of the haploid number (n) beyond diploid (2n), such as tetraploid (4n), hexaploid (6n), etc. The progeny of crosses between different ploidy levels typically exhibit substantial sterility

Fig. 5.3 Chromosome constitutions (karyotypes) in (a) diploid (2n = 22) and (b) tetraploid (2n = 44) button wrinklewort daisies [from Fig. 3 in Murray, B., Young, A.G. (2001). Widespread Chromosome Variation in the Endangered Grassland Forb *Rutidosis leptorrhynchoides* F. Muell. (Asteraceae: Gnaphalieae). Annals of Botany, 87(1), 83–90. Republished with permission of Oxford University Press].

(a) Diploid

1 2 3 4 5 6 7 8 9 10 11 12 13 14 15 16 17 18 19 20 21 22

(b) Tetraploid

1 2 3 4 5 6 7 8 9 10 11 12 13 14 15 16 17 18 19 20 21 22

23 24 25 26 27 28 29 30 31 32 33 34 35 36 37 38 39 40 41 42 43 44

Chromosomal differences can be detected using cytological methods

Cytological methods for chromosomal analyses in animals and plants are reviewed in references in the FURTHER READING for this chapter. Searching the web for "karyo-typing services" may identify commercial or research laboratories that will attempt to karyotype species for a fee or collaboration. Several organizations that archive living biological material are developing capacity for cytogenetics, including Frozen Ark and CryoArks in the UK, and the Australian Frozen Zoo.

Mechanism 2: Adaptation to different environments

There is now compelling evidence from many species that rapid development of repro-ductive isolation evolves primarily as a secondary consequence of genetic adaptation to different environments. In brief, positive associations between ecological divergence and reproductive isolation have been reported from field studies for > 500 species across several major taxa. For example, stickleback fish (*Gasterosteus* spp.) from three isolated lakes in British Columbia, Canada that had independently evolved benthic and limnetic forms showed low spawning rates when these forms were crossed, but normal rates in crosses between the same forms, both within and across lakes.

Whether adaptive differentiation is associated with all cases of reproductive isolation remains controversial, but adaptation is involved in most cases where it has evolved rap-idly in isolated populations (excluding those with chromosomal differences).

Mechanism 3: Coadaptive mechanisms

Coadapted gene complexes are combinations of alleles at different genes whose overall fitness effects are better than expected from the average effects of the constituent alleles. However, they are harmful when alleles involved in different coadapted complexes are recombined following crossing. For example, different combinations of mitochondrial and nuclear DNA genotypes in different populations have been shown to result in outbreeding depression following crossing, as observed in *Drosophila*, copepods, wasps (*Nasonia*), newts (*Triturus*), nematodes (*Caenorhabditis briggsae*), plants, and yeast (*Saccharomyces*). Such coadaptations are more likely between populations adapting to different, rather than similar environments, as found in an Australian bird (Morales et al. 2018).

> Coadapted gene complexes may develop in isolated populations within the same environment and result in outbreeding depression when the populations are crossed, but this is more likely for populations adapted to different environments

Empirical data on speciation in similar environments and genetic rescue meta-analyses both indicate that outbreeding depression in crosses between populations adapted to similar environments is rare, and if present, of a lesser magnitude than genetic rescue effects.

Predicting the risk of outbreeding depression

We used the above insights to predict the risk of outbreeding depression. There is a low risk of outbreeding depression for crosses of populations with no fixed chromosomal differences, that are adapted to similar environments, and were formerly connected by gene flow. For example, the old-field mouse sub-species and the populations of golden lion tamarins have these characteristics, and their crosses do not show outbreeding depression (Table 5.5).

Conversely, if crosses are made between species or populations differing by fixed chromosomal differences, or populations adapted to different environmental conditions, especially those separated for long periods of times, harm may result. Species crosses are often harmful—horses crossed with donkeys produce sterile mules, because these species have multiple fixed chromosomal differences, have been isolated for ~ 2 million years, and have evolved on different continents. Crosses within species may be harmful, if the crossed populations are sufficiently distinct. For example, negative effects were seen when differentially adapted partridge pea populations in the USA were crossed—F_1 crosses were beneficial, but F_3 crosses at distances of $\geq 1{,}000$ km were harmful, although these recovered in later generations (see next section). However, even crosses between species are not always harmful, and some are beneficial: what matters is the presence and extent of the mechanisms outlined above. We address this issue in more detail in Chapter 7.

Partridge pea (USA)

Populations usually recover from outbreeding depression over generations

Crossed populations that exhibit outbreeding depression will usually recover their fitness over subsequent generations due to natural selection acting upon the extensive genetic variation in the hybrid population. Such recovery has been observed in all reported cases of which we are aware, both for crosses between species and for crosses between locally adapted populations within species. For example, three sunflower species (*Helianthus anomalus*, *H. deserticola*, and *H. paradoxus*) have each evolved from crosses between the same two parents *H. annuus* and *H. petiolaris* that differ by 10 fixed chromosomal rearrangements. F_1 crosses between the two parent species have pollen viability of $< 5\%$ and 2.1% viable seed, but experimental crosses between them exhibited recovery of pollen viability to $> 90\%$ over five generations as a result of natural selection. In crosses between distant populations of partridge peas that showed strong F_3 hybrid breakdown, fitness had improved to a level superior to the parents by the F_6 generation.

Sometimes the crossed population attained higher eventual fitness than either parental population. However, if population size is very small and outbreeding depression severe, species with low fecundity may go extinct before crosses can recover from outbreeding depression.

Conclusion

Genetic rescue should be routinely considered as a conservation option to promote long-term recovery for small outbreeding populations with limited gene flow. Given the available evidence, the current reluctance to attempt genetic rescues in conservation settings is not scientifically justified. We recommend a much broader use of gene flow into small inbred populations to enhance fitness, population persistence, and ability to evolve, and ultimately to reduce species extinctions.

Summary

1. Crossing of an inbred population to an unrelated population reduces inbreeding and increases genetic diversity.
2. Crossing may have beneficial or harmful effects on fitness, but beneficial effects are more common, and the circumstances associated with harmful effects are understood and predictable.
3. The primary mechanisms causing outbreeding depression in crosses between populations are fixed chromosomal differences and adaptive genetic differences, especially for populations isolated for many generations. Even if outbreeding depression occurs, it is often only temporary, as natural selection removes it, especially in larger populations.
4. When the risk of outbreeding depression is low, crossing an isolated inbred population to another population has overwhelmingly beneficial effects, with a median increase of 148% under stressful conditions and 45% under benign conditions.
5. The benefits of genetic rescues typically persist across generations in outbreeding species unless population size remains small.
6. Gene flow into populations with low genetic diversity also strongly benefits their ability to evolve.
7. We recommend instituting gene flow between populations of outbreeding species to reverse the detrimental effects of inbreeding and loss of genetic diversity when the proposed crosses have a low risk of outbreeding depression and the benefits justify the cost of genetic rescue.

FURTHER READING

Erickson & Fenster (2006) A cross between distant populations within partridge peas recovered in fitness from F_3 outbreeding depression to fitness superior to the parents in the F_6 generation.

Frankham (2015, 2016, 2018) These meta-analyses revealed that gene flow into small inbred populations results in large and consistent benefits on fitness and ability to evolve, and that the fitness benefits persist across generations in naturally outbreeding species.

Frankham et al. (2011) Reviewed mechanisms leading to the evolution of outbreeding depression and proposed means to predict its occurrence.

Frankham et al. (2017) *Genetic Management of Fragmented Animal and Plant Populations*: Chapters 6 and 7 have more detailed treatments of topics in this chapter.

Johnson et al. (2010) Describes the beneficial fitness effects of crossing a small inbred population of Florida panthers to Texas individuals.

Kirov et al. (2014) Describes cytological methods for use in plants.

Rowell et al. (2011) and Houck et al. (2017) Describe cytological methods for use in animals.

Rundle et al. (2000) Classic study on the role of adaptive differentiation in outbreeding depression, based on studies in stickleback fish.

SECTION II
Making genetic management decisions

Having described genetic problems associated with reduced gene flow and identified restoration of gene flow as a remedy, we are now ready to consider the genetic management of fragmented populations. Critically, we have identified the important risk factors for outbreeding depression that allow managers to minimize the chance of experiencing this problem.

Our objective in genetic management of threatened species is to minimize the risk of population and species extinction due to genetic factors and their interactions with other risk factors.

The following questions need to be answered to achieve this.

Is the taxonomy reliable and appropriate for conservation purposes?

We provide guidelines for determining whether current taxonomy is adequate for conservation purposes (Chapter 6). If not, we recommend appropriate definitions of species and means to implement appropriate species delineations with them.

Are there isolated populations suffering genetic erosion?

The second question is whether there are populations with low genetic diversity that are inbred and have an elevated risk of extinction. If so, we next ask (Chapter 7):

- Would they benefit from gene flow?
- If yes, are there any populations that can be used to genetically rescue them?
- If yes, would crossing between populations be beneficial or harmful?
- If beneficial, would the cross yield sufficient benefits to justify resourcing a genetic rescue attempt?

If yes, we turn to the details of managing gene flow.

A Practical Guide for Genetic Management of Fragmented Animal and Plant Populations. R. Frankham, J. D. Ballou, K. Ralls, M. D. B. Eldridge, M. R. Dudash, C. B. Fenster, R. C. Lacy & P. Sunnucks. Oxford University Press (2019). © R. Frankham, J. D. Ballou, K. Ralls, M. D. B. Eldridge, M. R. Dudash, C. B. Fenster, R. C. Lacy & P. Sunnucks 2019. DOI: 10.1093/oso/9780198783411.001.0001

How can we manage gene flow?

If gene flow between populations is likely to yield worthwhile benefits, we consider the choice among donor populations, and how much gene flow from the donor(s) into the isolated inbred population is desirable (Chapter 8). Worthwhile genetic management of fragmented populations can usually be conducted with minimal information on the populations or information obtained at minimal cost. However, management efficiency can typically be improved with more detailed information, particularly by using mean kinship. However, we emphasize that **any augmentation of gene flow is nearly always better than none** when the risk of outbreeding depression is low.

How should we modify genetic management under global climate change?

In Chapter 9 we address genetic management of fragmented populations under global climate change, applying the principles we developed earlier to the even more trying situation of a persistently changing environment. Global climate change increases the need for genetic management and its urgent implementation where necessary, because genetic diversity is necessary for populations to adapt to changing conditions.

We end by emphasizing the need to integrate genetic management with other demographic and ecological considerations across populations, disciplines, institutions, and political boundaries to develop a comprehensive plan for the conservation of a species, such as the "One Plan Approach." Details of that approach are beyond the scope of our book. However, there may be the need to compromise on favored genetic management options to address other competing demands.

Conservation managers we consulted wanted to know not only what to do but also **what not to do**. Accordingly, we conclude this section by listing inappropriate genetic management actions (Table SII.1).

Table SII.1 What not to do in genetic management.

Management action	Adverse consequences
Failing to consider the consequences of doing nothing	Populations may unnecessarily go extinct. **This may be the riskiest option, rather than the most cautious**
Failure to review taxonomy to see if it is appropriate	May result in outbreeding depression or preclusion of genetic rescues
Accepting taxonomy based on the Phylogenetic or General Lineage Species Concepts for allopatric populations	Over-splitting of taxa likely, meaning that genetic rescues may be precluded because populations have been designated as separate taxa

Management action	Adverse consequences
Automatically concluding that genetically distinguishable populations should be separately managed without evaluating the desirability of genetic rescues	Extinction of small populations due to genetic and combined genetic and non-genetic factors
Failing to manage across political boundaries crossed by a species distribution	Inadequate genetic management, likely resulting in an elevated risk of extinction, compared to combined management
Crossing populations without assessing the risk of outbreeding depression	Higher risk of outbreeding depression than necessary
Crossing populations with different ploidies	Sterile offspring, lost resources
Failing to consider the consequences of self-fertilizing mating system	Inappropriate genetic management may result, such as lower initial fitness benefit of genetic rescue and reduced persistence of benefit over generations
Basing choice of populations to use in genetic rescues on F_{ST}	Risk of gene flow from inappropriate populations with low genetic diversity, and poorer outcomes than when donor populations are chosen based on minimizing mean kinship
Selecting against a genetic defect without considering the overall genomic consequences	Selection may remove many individuals from the breeding pool, resulting in faster decline in genetic diversity and more inbreeding than without the selection, thereby increasing extinction risk
Managing populations to conserve rare alleles (often those in the MHC or microsatellites)	Higher rates of inbreeding and lower levels of genome-wide genetic diversity than achievable by managing using mean kinship, and elevated levels of harmful alleles
Not breeding from taxon hybrids, such as sub-species that yield fertile offspring with normal survival	Valuable genetic material that may aid in coping with global climate change through *in situ* evolution may not be utilized. Taxonomic revision may later classify the parent populations as not being distinct.

Appropriate delineation of species for conservation purposes

Delineating species boundaries is a necessary first step before decisions on the management of populations can be made. However, there are many problems with species delineation, including diverse species definitions, a lack of standardized protocols, and poor repeatability. Definitions that are too broad will lead to out-breeding depression if populations are crossed, while those that split excessively may preclude genetic rescue of small inbred populations with low genetic diversity. To minimize these problems, we recommend the use of species concepts based upon reproductive isolation and advise against the use of Phylogenetic, General Lineage, and Taxonomic Species Concepts. We provide guidelines as to when taxonomy requires revision and outline protocols for robust species delineations.

TERMS

Allopatric, Biological Species Concept, chloroplast DNA, conspecific, fixed gene differences, General Lineage Species Concept, gene tree, lineage sorting, parapatric, Phylogenetic Species Concept, phylogenetic tree, reciprocal monophyly, sympatric, Taxonomic Species Concept

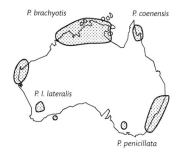

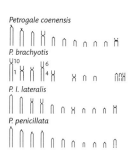

The taxonomy of rock-wallabies in Australia was controversial, but has been resolved using extensive geographic sampling, chromosomal, molecular genetic, and morphological analyses, as well as crosses between candidate taxa. One species is illustrated, along with the distributions and chromosomes for four other species (Frankham et al. 2010, p. 362). The chromosomes with numbers represent centric fusions.

A Practical Guide for Genetic Management of Fragmented Animal and Plant Populations. R. Frankham, J. D. Ballou, K. Ralls, M. D. B. Eldridge, M. R. Dudash, C. B. Fenster, R. C. Lacy & P. Sunnucks. Oxford University Press (2019). © R. Frankham, J. D. Ballou, K. Ralls, M. D. B. Eldridge, M. R. Dudash, C. B. Fenster, R. C. Lacy & P. Sunnucks 2019. DOI: 10.1093/oso/9780198783411.001.0001

Appropriately resolved taxonomy is crucial to conservation management

If populations and species are not appropriately delineated, genetically differentiated populations (including distinct species) may be inappropriately crossed, leading to out-breeding depression. Alternatively, if populations that are inbred and depleted of genetic diversity are inappropriately classified as distinct taxa, this may impede genetic rescue through regulatory or legal hurdles.

In what follows, we provide guidance to assist conservation managers in deciding whether the current taxonomy of their species is adequate or needs re-evaluating.

What is our objective?

We seek to have species delineated such that their probability of persistence is maximized

The probability of species persistence will be maximized if they have high fitness, ample genetic diversity to evolve, and large population sizes (Chapters 2–5 and 9). Thus, species delineations for conservation purposes need to facilitate achievement of these conditions.

As species are meant to be genetically differentiated, we now consider how such population differences are characterized.

Characterizing genetic differentiation in taxonomy

Genetic differentiation of populations can be characterized in terms of reproductive isolation, or indirectly as differentiation for heritable markers (indicating reproductive or physical isolation)

If populations are in contact but exhibit no gene flow, they are reproductively isolated. Genetic markers, chromosomes, or heritable morphological differences can provide evidence of isolation. Deliberate crossing attempts provide ideal evidence, if feasible.

If populations have non-overlapping differences in morphology, behavior, or life history they are often delineated as distinct species. Since such characters may differ in response to diverse environmental conditions, rather than representing genetic differences, taxonomists increasingly make use of "near neutral" molecular genetic markers, such as microsatellites, SNPs, and DNA sequence data.

Completely diverged populations are diagnosed by having one or more of the following: non-overlapping distributions for heritable quantitative characters, no shared alleles at variable markers, homozygosity for different alleles, or being reciprocally monophyletic in phylogenetic trees (defined below)

Isolated populations derived from the same ancestral population drift apart and eventually become homozygous, some for the same ancestral allele and some for different ancestral alleles (lineage sorting) or less frequently for new mutations. In taxonomy, diverged populations are described as having no shared alleles at polymorphic markers, being homozygous for different alleles (fixed gene differences) at some markers, or exhibiting reciprocal monophyly in phylogenetic trees. Reciprocal monophyly occurs when the members of a group are evolutionarily more closely related to each other than they are to any members of an alternative group: this is the basis of species delineation according to some species concepts.

Taxonomists frequently represent evolutionary similarities among groups of individuals in phylogenetic trees, where increasing levels of differentiation progress from a star phylogeny, through polyphyly, paraphyly, to reciprocal monophyly (Fig. 6.1).

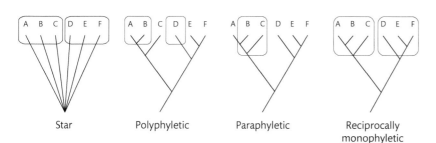

Fig. 6.1 Phylogenetic trees illustrating increasing divergence of populations from left to right: star phylogeny (all descendant populations are similarly related to recent ancestral population), polyphyly (ABD versus rest), paraphyly (BC versus rest), and reciprocal monophyly (ABC versus DEF) (Frankham et al. 2017, Fig. 9.2).

Critically, **gene trees inferred for the same populations using different genetic markers are often discordant due to genetic drift.** Consequently, only combined information from multiple genetic markers is likely to yield a reliable phylogenetic tree. Thus, **we recommend use of data from multiple independent genetic markers in species delineations.**

This chapter defines problems with delineating species and recommends methods for avoiding these issues in delineations for conservation purposes.

There are many problems with species delineations

Species delineation as currently practiced is often deeply flawed, with serious consequences for conservation of biodiversity (Garnett & Christidis 2017). The main problems are:

- no standardized sampling regimes, characters, or analyses
- widespread use of diverse methods
- instability of delineations to technological change
- many disparate definitions of species that often lead to different delineations
- poor repeatability of delineations
- over-lumping is common (especially in older delineations)
- over-splitting is common currently, and worsening
- widespread use of markers that have little statistical robustness (e.g. sole reliance on mitochondrial DNA [mtDNA] or chloroplast DNA [cpDNA])
- defining morphologically similar plant populations with different chromosomal numbers (ploidy levels) as belonging to the same species (conspecific), when crosses result in sterility
- failure to check chromosomes, meaning that chromosomally caused reproductive isolation may be missed
- rarely evaluating whether morphological differences used in taxonomic delineations reflect heritable versus environmental differences (e.g. in common garden experiments).

Several of the above issues are illustrated in Box 6.1 by the controversial and changing taxonomy of the Bornean and Sumatran orangutans (*Pongo pygmaeus* and *P. abelii*).

Box 6.1 Controversial and changing taxonomy of recently diverged Bornean and Sumatran orangutans, based on conflicting information and different species concepts

(after Frankham et al. 2017, Box 9.1)

Orangutans on the islands of Borneo and Sumatra in Southeast Asia have been designated as sub-species or separate species, based on possible differences in morphology, behavior, chromosomes, mtDNA sequences, and nuclear genes. Since they differed genetically by at least as much as chimpanzees (*Pan troglodytes*) and bonobos (*Pan paniscus*), and their divergence was dated at 10.5 million years using mtDNA, full species status for the two forms was suggested.

However, hybrids between Bornean and Sumatran orangutans are viable and fertile in the F_1 and F_2 generations, meaning they are not distinct species according to the Biological Species Concept. Further, a broader geographic sampling of individuals revealed that the two forms do not have fixed differences in mtDNA, and new molecular dating based on whole genome sequences yielded a divergence time of only 334,000 years. However, fixed differences for nuclear genetic markers and for a chromosomal inversion are sufficient to designate the two forms as separate species under the Phylogenetic Species Concept (see later), and recently, a third species of orangutans has been delineated by proponents of this concept (Nater et al. 2017). This illustrates the confusion created by use of different characters, inadequate sampling regimes, and different species definitions.

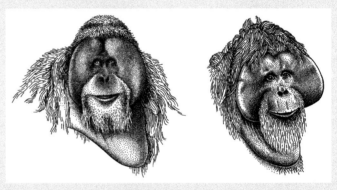

Bornean and Sumatran orangutans

In what follows, we seek to minimize these problems by advocating:

- appropriate species concepts for conservation purposes
- use of multiple characters and genetic markers (integrative taxonomy)
- wide sampling of species distributions
- use of scientifically robust analyses
- development of standardized protocols for robust species delineations in conservation contexts.

We concentrate on defining appropriate species concepts for conservation purposes and defer most considerations of the remaining issues to Appendix 3. Box 6.2 provides examples of what we consider thorough versus inadequate taxonomic determinations.

Box 6.2 Thorough versus inadequate taxonomic delineations

(after Frankham et al. 2017, Box 9.2)

Thorough taxonomic delineations: extensive sampling and use of multiple lines of evidence

Thorough taxonomic delineations are typically based on wide geographic sampling of individuals along with evidence from multiple characters, as illustrated by rock-wallabies (*Petrogale*) in Australia (Chapter frontispiece). Rock-wallabies are small macropodid marsupials < 1 m tall, with fragmented distributions on rocky outcrops, and they have diverged relatively recently. Their taxonomy was controversial, with between 5 and 11 species (many threatened) recognized by different authors. In 1976 studies were begun at Macquarie University to resolve their taxonomy using extensive geographic sampling, chromosomal analyses, allozymes, mtDNA, DNA–DNA hybridization, nuclear DNA sequences, morphology, and crosses between candidate taxa. These studies revealed 20 sharply delineated chromosomal races (four shown on the frontispiece map) whose distributions were typically concordant with one or more other attributes (morphology, results of crossing, nuclear DNA, or mtDNA). The 20 races are now recognized as distinct taxa.

Inadequate: diploid and polyploid forms within the same "species"

Endangered button wrinklewort daisies (*Rutidosis leptorrhynchoides*) in southeastern Australia contain morphologically similar diploid (2n), tetraploid (4n), and hexaploid (6n) chromosomal forms that when crossed result in outbreeding depression. These forms should be designated as distinct species for conservation purposes and managed separately. The inclusion of different ploidy levels within a single named species is a widespread problem in plant taxonomy.

Button wrinklewort daisy (Australia)

There are ~ 30 species concepts, which often result in different species delineations

Table 6.1 defines the four concepts most widely used by the systematic and conservation communities: the Biological Species Concept, the General Lineage Species Concept, the Phylogenetic Species Concept, and the Taxonomic Species Concept. Worryingly, many species delineations or revisions fail to specify what species concept has been used and rely upon the judgment of taxonomists (referred to as the Taxonomic Species Concept).

Table 6.1 Species definitions according to different concepts.

Species concept	Species definition	Reference
Biological (BSC)	"groups of actually or potentially interbreeding natural populations, which are reproductively isolated from other such groups"	Mayr (1942)
General Lineage (GLSC)	"species are (segments of) separately evolving metapopulation lineages"	de Queiroz (2007)
Phylogenetic (PSC)	"a species is the smallest diagnosable cluster of individual organisms within which there is a parental pattern of ancestry and descent"	Cracraft (1983)
Taxonomic (TSC)	"whatever a competent taxonomist chooses to call a species"	Wilkins (2009)

Despite the disparate definitions, species concepts typically indicate that species are cohesive clusters of individuals that have at least partially different evolutionary paths and represent different lineages. All serious concepts recognize that populations inherently incapable of gene exchange are distinct species, while those that can freely exchange genes when in contact belong to the same species. However, there are major differences in the treatment of fragmented (allopatric) populations capable of gene flow without adverse fitness consequences or with beneficial consequences.

Use of PSC results in more splitting than BSC. For example, PSC yielded 49% more species than BSC when used to delineate the same diverse group of organisms, and revisions using PSC approximately doubled the number of primate and ungulate species.

Changed taxonomy due to the application of different species concepts can lead to altered or confused management, with adverse consequences for biodiversity conservation, as occurred for example with black and white ruffed lemurs (*Varecia variegata*) and orangutans (Box 6.1).

Species delineations are inappropriate if they create groups that are too inclusive (over-lumping) or too small (over-splitting). How do these problems arise, and how can they be minimized?

Over-lumping arises primarily from the use of morphological characters with insufficient resolving power

Many "species" contain reproductively isolated segments that have subsequently been delineated as separate species. Numerous species were delineated hundreds of years ago, based on limited information and sampling, both numerically and geographically, with insufficient resolving power. For example, combined molecular genetic, chromosomal, and morphological studies show that Australia is home to well over 100 locally distributed species of velvet worms (Onychophorans), rather than the seven widespread morphological species previously recognized. Further, cryptic adaptively differentiated

Black and white ruffed lemur (Madagascar)

lineages of fluted gum (*Eucalyptus salubris*) exist in southwestern Australia and show no evidence of gene flow in areas of geographic overlap.

Issues of resolving power can usually be overcome by use of multiple molecular markers or genome sequencing plus chromosomal assays, especially when combined with other characters.

Over-splitting can arise from the use of PSC or GLSC on small allopatric populations

In small isolated allopatric populations, genetic drift and mutation will lead to units that are diagnosably different at the molecular and perhaps the phenotypic level, although they are not intrinsically reproductively isolated and may be ephemeral under historical patterns of population separation and re-connection. This is a major issue in the conservation of fragmented populations, because **the very measure that is used to delineate PSC and GLSC species (genetic differentiation among populations) is causally related to low genetic diversity within small populations** (Chapters 4 and 7). However, this distinctiveness does not necessarily indicate adaptation to different environments but can reflect loss of alleles in the inbred population due to genetic drift. For example, the Illinois population of the greater prairie chicken was once connected to populations in nearby states, but in historical times became a small and isolated genetic outlier (Fig. 6.2), yet it is not reproductively isolated from the other populations: crossing between populations enhanced fitness.

Greater prairie chicken (USA)

Fig. 6.2   The small Illinois (IL) population of greater prairie chickens is an outlier based on microsatellite data, compared to larger Nebraska (NE), Kansas (KS), and Minnesota (MN) populations, but is not reproductively isolated from them. Map of the geographic distribution of the greater prairie chicken showing the location of studied populations. Estimated census population sizes are Illinois < 50; Kansas > 100,000; Minnesota > 4,000; Nebraska > 100,000 (Frankham et al. 2017, Fig. 9.3, based on Bouzat et al. 1998, Figs 1 and 3).

Threatened populations maintained for the number of generations required to exhibit fixed gene differences or reciprocal monophyly are at considerable risk of extinction from inbreeding and are expected to exhibit large genetic rescue effects upon crossing. These expectations have been verified in empirical studies.

PSC and GLSC are also unstable under technological change. Technological advances from analyses of morphology to allozymes, microsatellites, SNPs, and now whole genome sequencing have led to finer and finer diagnosable differences among populations. These differences are considered grounds for naming new species using PSC and GLSC. This approach has created instability of species delineations, rather than improving certainty of delineations with increasing information.

> Use of mtDNA as the sole basis for delineating species is likely to result in over-splitting and taxonomic distortions due to the impacts of selection

Small diverged and isolated populations are especially susceptible to being classified as different species according to PSC, or GLSC, when mtDNA is used as the sole basis for delineations. The problems inherent in using mtDNA include that it:

- has very low precision for estimating phylogenetic trees, as it is a single inherited unit whose tree may differ from those determined for nuclear genes, and from the "true" species tree, as for example was observed in silvereye birds (*Zosterops*) in Australia
- is subject to distortions due to selection when a new favorable mutation in a single base rises to fixation and in the process sweeps away genetic diversity throughout the whole molecule, as it is a non-recombining unit in most species.

As mtDNA diverges more rapidly than nuclear diploid loci (due to lower N_e and higher mutation rates), it has been widely used in taxonomy. Similar issues apply to the use of chloroplast DNA (cpDNA). Consequently, **we are opposed to the sole use of mtDNA or cpDNA data for taxonomic delineations due to their low precision and high potential for misleading inferences.**

Species definitions used in conservation should not compromise species persistence

> Species concepts used in conservation should contribute positively to population and species persistence, by creating designations that simultaneously minimize outbreeding depression if populations are crossed, while allowing for genetic rescue of small inbred populations

A hypothetical example illustrates how both these objectives can be achieved simultaneously by using appropriate species concepts. The consequences of different species delineations for six hypothetical populations are illustrated in Fig. 6.3. Too broad a delineation of species in case 1 (as might occur using only morphological data) leads to a high risk of outbreeding depression when populations **a** and **b** cross. Over-splitting in case 2, because of large genetic drift effects in small populations, leads the small **a4** population to be classified as a distinct species under PSC and GLSC, with no other populations within its species available to rescue it genetically or reinforce it demographically. In case 3, use of reproductive isolation (defined to include deleterious consequences of crossing on reproduction or survival) to delineate species **a** versus species **b** both minimizes the risk of outbreeding depression and allows genetic rescue of small inbred populations within species.

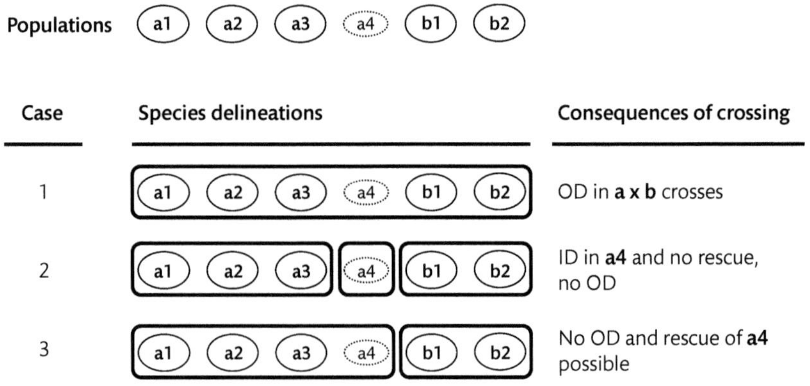

Fig. 6.3 Consequences of crossing populations following different species delineations in relation to outbreeding depression (OD), inbreeding depression (ID), and genetic rescue. Populations **a** and **b** are reproductively isolated (show OD on crossing), but populations within them reproduce freely. The small isolated **a4** population has low N_e, is inbred, has low genetic diversity, and shows high divergence from the other populations (Frankham et al. 2012, Fig. 1).

The preceding arguments lead us to recommend that substantial intrinsic reproductive isolation be used to define species of outbreeding sexual organisms for conservation purposes. BSC captures large components of what is needed for conservation, especially when used in a "relaxed" form that accepts low rates of gene flow between species.

We do not recommend use of PSC and GLSC as they are prone to over-splitting and yet may miss chromosomally caused reproductive isolation. Such over-splitting, sometimes in an attempt to promote greater conservation of biodiversity, can preclude augmentation of gene flow to preserve taxa with small population sizes, cause limited resources to be spread more thinly, and thereby result in greater loss of biodiversity. **We are opposed to the use of TSC**, because it does not allow managers to make knowledgeable decisions on whether to cross populations or to manage them separately.

How much differentiation is required to classify populations as distinct species?

Cut-offs for delineating species should be based upon those for well-recognized species defined using intrinsic reproductive isolation

Since reproductively isolated species arise by relatively gradual genetic divergence as envisaged by Darwin (apart from more or less instantaneous origins of polyploids), there is a continuum from random mating to complete reproductive isolation. Cut-offs for species lie at the higher end of the gradient of reproductive isolation, but requiring total reproductive isolation is too extreme, as many well-recognized species exhibit low residual levels of gene flow with other species (vonHoldt et al. 2018). When delineating species in the absence of information on interbreeding, following cut-offs used for well-recognized species defined using the BSC or related species concept provides useful guidelines.

Conclusions about appropriate species concepts depend on the geographic distribution of populations

Problems inherent in using different species concepts for delineating species for conservation purposes occur with allopatric distributions, rather than parapatric, or sympatric ones (Fig. 6.4).

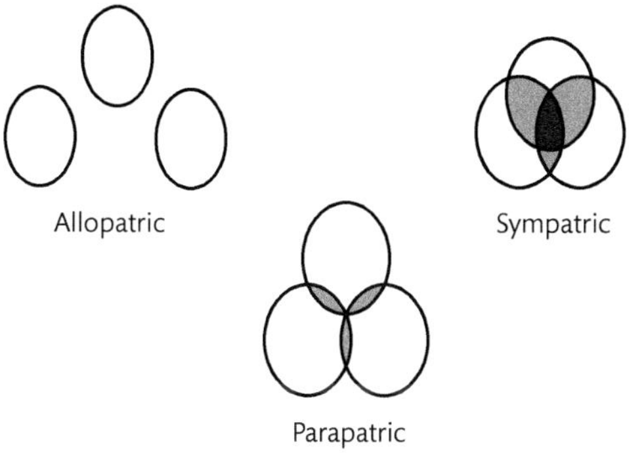

> Species concepts based upon intrinsic reproductive isolation are recommended for conservation purposes

Allopatric

Sympatric

Parapatric

Fig. 6.4 Allopatric, sympatric, and parapatric distributions of populations (Frankham et al. 2017, Fig. 9.6).

Velvet worm

For populations with sympatric and parapatric distributions, all major species concepts lead to similar species delineations

Since sympatric and parapatric populations have at least partially overlapping distributions, they have the opportunity to mate. Therefore, a lack of shared alleles at one or more autosomal loci (given appropriate sampling) is sufficient to establish intrinsic reproductive isolation between populations. Thus, they will be classified as separate species under all major species concepts (even evidence of very limited gene flow should be acceptable). For example, samples of outcrossing velvet worms (Onychophora) from the same log in the Blue Mountains west of Sydney, Australia showed fixed gene differences at 70% of allozyme genes and were reclassified as distinct genera.

For allopatric populations, different species concepts often lead to different species delineations

Allopatric populations will usually be classified appropriately using BSC. However, allopatrically distributed populations delineated using diagnostic PSC or GLSC are often not intrinsically reproductively isolated. However, for many allopatric populations we do not have information on reproductive isolation from crossing experiments through to the F_3 generation. Fortunately, fixed chromosomal differences and adaptation to strongly different environments can be used to predict reproductive isolation (Chapter 5).

The degree of reproductive isolation required to delineate allopatric species can be guided by examining trait differences previously used for sympatric and parapatric species (that are by definition showing a degree of reproductive isolation). Trait differences among populations for morphological and song characteristics in birds that justify distinct species status have been suggested based on this approach.

How do we decide whether a taxonomic revision is required?

Taxonomic revisions will often be required for species with allopatric distributions delineated using PSC or GLSC

If the taxonomy of a species with an allopatric distribution was determined purely on morphology, cpDNA, or mtDNA, on limited sampling, or on use of PSC, GLSC, or TSC, it likely requires revision.

Conversely, taxonomic delineations based on crossing results, or congruent data from multiple characters (chromosomes, several independent genetic markers, morphology, etc.) on geographically well-sampled distributions, and where BSC has been used in the delineations, should be adequate. Birds are less likely to require taxonomic revision, as their delineations are usually done using a well-organized system by Birdlife International, based upon use of BSC.

For sympatrically or parapatrically distributed populations, species delineations conducted under any of the species concepts should be reliable and repeatable, given adequate sampling regimes and sample sizes, and they were not based solely on mtDNA or cpDNA data.

How should a taxonomic re-evaluation be conducted if required?

The principles and issues we have discussed above provide a basis for species delineations for conservation purposes. Further details are provided in Appendix 3. We endorse the trend towards use of multiple lines of evidence in species delineations, as in integrative taxonomy. However, we recommend that the evidence should always include chromosomal data, and the delineation be based on observed or predicted reproductive isolation.

What should we do about genetically differentiated populations within species?

Some populations within species may be sufficiently distinct to justify separate management

As there is a continuum of levels of genetic differentiation and reproductive isolation among populations, some populations within species justify separate management. We recommend that the same principles we have applied above to species delineations be used to delineate units within species for genetic management purposes, but by using less extreme trait differences in delineations (i.e. allowing for modest reproductive isolation or outbreeding depression).

When there are already identified units within a species, such as sub-species and evolutionarily significant units (ESUs), they cannot be relied upon if they suffer from similar shortcomings to those we identified for species, including a diversity of definitions. Previously identified units within species will often need to be revised, as they frequently suffer from these shortcomings.

Summary

1. A critical first step in conservation is to define species for conservation purposes, but species delineations have often been done inappropriately or in a scientifically unsatisfactory manner.
2. Species delineations that are too broad will often lead to outbreeding depression when populations are crossed, while those that split excessively may preclude genetic rescue of small inbred populations with low genetic diversity.
3. Use of reproductive isolation to define distinct species minimizes the risk of outbreeding depression, while retaining the option for genetic rescue of populations suffering genetic erosion. Consequently, we recommend the use of the Biological Species Concept and related concepts when delineating species for conservation purposes.
4. Conversely, use of the diagnostic Phylogenetic, General Lineage or Taxonomic Species Concepts is not recommended, as they will often lead to excessive splitting for allopatric populations.

5. For sympatric or parapatric populations, distinct species are diagnosed by any genetically based distinctiveness that indicates lack of (or very limited) gene flow, and different species concepts typically yield concordant delineations.

6. For allopatric populations, crossing data are ideal, but if they are unavailable, reproductive isolation can be inferred from fixed chromosomal differences, and/or adaptive differentiation among populations, allowing delineation under BSC.

7. Reliable and scientifically sound species delineations require adequate geographic sampling, data on multiple characters (chromosomes, multiple independent genetic markers, morphology, life history, ecology, etc.), and the use of BSC. When these conditions have not been satisfied, taxonomic revisions are typically required, especially for species with allopatric distributions.

8. Some populations within a species may need separate management, but existing sub-specific units, such as sub-species and ESUs, suffer from the same problems as species and will often need to be revised.

FURTHER READING

Coates et al. (2018) Excellent review of taxonomy and conservation in the age of genomics.

Coyne & Orr (2004) *Speciation*: Outstanding textbook on speciation and its causes.

Frankham et al. (2012) Evaluated the implications of different species concepts for conserving biodiversity, and recommended concepts based on reproductive isolation.

Frankham et al. (2017) *Genetic Management of Fragmented Animal and Plant Populations*: Chapter 9 has a more detailed treatment of topics in this chapter.

Hall (2017) *Phylogenetic Trees Made Easy: A How-to Manual*: A simple introduction to the construction of phylogenetic trees from genetic data.

Rieseberg & Willis (2007) Authoritative review on plant speciation.

SOFTWARE

IQ-TREE: reputed to be very easy to use, fast, and to give phylogenetic trees very similar to other reputable software (Nguyen et al. 2015). http://www.iqtree.org

Phylogeny.fr: robust phylogenetic analysis for the non-specialist (Dereeper et al. 2008). http://www.phylogeny.fr/

Phylogeny Programs: a compilation of most of the software programs for phylogenetic analysis, compiled by Joe Felsenstein at the University of Washington. http://evolution.genetics.washington.edu/phylip/software.html

Simple Phylogeny: a simple program for producing phylogenetic trees from genetic data (Larkin et al. 2007). https://www.ebi.ac.uk/seqdb/confluence/display/THD/Simple+Phylogeny

TreeBASE: phylogenetic database of DNA sequences among species and phylogenies derived from them (Piel et al. 2009). www.treebase.org

Are there populations suffering genetic erosion that would benefit from gene flow?

CHAPTER 7

The current response when populations are genetically differentiated is usually to recommend keeping the populations isolated and instituting separate management. This is often ill-advised. A paradigm shift is needed where evidence of genetic differentiation is followed by an assessment of whether fragments are suffering genetic erosion, and if so, whether there are other populations to which they could be crossed. If yes, we should next ask whether crossing is expected to be beneficial or harmful, and if beneficial, whether the benefits would be large enough to justify a genetic rescue attempt. Here we address these questions based on principles established in the preceding chapters.

TERMS

Monomorphic

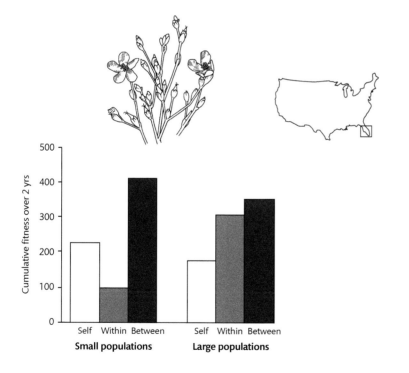

Experimental crossings of the naturally outbreeding highlands scrub hypericum (*Hypericum cumulicola*) in Florida produced different results in small and large populations. In the small populations, selfing did not result in inbreeding depression, within population crossing resulted in low fitness, and crossing with another population resulted in substantial genetic rescue for fitness. Conversely, in the large populations, selfing resulted in inbreeding depression, within population crosses had higher fitness than those in the small populations, and between population crosses showed much smaller increases in fitness than observed in the between population crosses for the small populations (after Frankham et al. 2017, p. 223, based on Dolan et al. 1999; Oakley & Winn 2012).

A Practical Guide for Genetic Management of Fragmented Animal and Plant Populations. R. Frankham, J. D. Ballou, K. Ralls, M. D. B. Eldridge, M. R. Dudash, C. B. Fenster, R. C. Lacy & P. Sunnucks. Oxford University Press (2019). © R. Frankham, J. D. Ballou, K. Ralls, M. D. B. Eldridge, M. R. Dudash, C. B. Fenster, R. C. Lacy & P. Sunnucks 2019. DOI: 10.1093/oso/9780198783411.001.0001

Separate management of genetically differentiated populations is often recommended

Evidence of population structure and limited gene flow frequently leads to the questionable conclusion that populations should be managed as separate units. For example, some authorities proposed that fragmented populations of Blanding's turtle (*Emydoidea blandingii*) in the Chicago area be managed as separate units (Box 7.1). Further, separate management was recommended for isolated populations of the endangered eastern bristlebird (*Dasyornis brachypterus*) in Australia, despite negligible genetic differentiation and some remnants having reduced genetic diversity. A preference for separate management probably explains why there have been only ~ 30 genetic rescues done for conservation purposes, yet more than a million isolated populations of threatened species would likely benefit from gene flow (Chapter 1).

Box 7.1 Should isolated, declining populations of Blanding's turtles in central North America be managed as separate populations?

(Rubin et al. 2001; King 2013; Illinois Natural History Survey 2016; Anthonysamy et al. 2018)

Blanding's turtles were once much more common across the northeastern USA and southeastern Canada, but are now sparsely distributed in highly fragmented populations and listed as globally endangered. Some biologists have argued for keeping separate geographically distant and long-separated populations, but even cautioning that turtles should perhaps not be moved among populations as close as 10 km apart in the Chicago area because this might disrupt locally adapted gene pools. This proposal derives from molecular genetic data showing different allele frequencies among some of the local breeding ponds. Most of the remnant populations in Illinois are in the counties surrounding the Chicago area, predominantly with N < 50 and thus N_e likely only ~ 10. However, the reduced genetic diversity relative to larger populations elsewhere and partial genetic divergence between some local populations are almost certainly due to genetic drift, rather than natural selection. The isolation of the populations likely occurred over the last 100 years (about three turtle generations) as woodland–prairie mix in the area was converted to Chicago's suburbs. Adaptive divergence is unlikely due to the short time the ponds have been isolated and the very small size of the populations. Given the human-induced environmental changes, even in the unlikely event that there were pre-existing, fine-scale adaptive differences, they would be to environments that no longer exist. In such a scenario, without good reasons to the contrary, the risks of inbreeding depression and demographic problems far outweigh the risks of loss of important local distinctiveness. It would be preferable to manage the turtles within geographic regions such as the Chicago suburbs and surrounding counties as one, not multiple units. Managers are now evaluating translocation options to protect genetic diversity and avoid inbreeding in the local breeding populations.

Blanding's turtle (North America)

Separate management based on genetic distinctiveness is a recipe for extinction of many fragmented populations because the population fragments that are most highly differentiated from other population fragments are usually those with the lowest genetic diversity. This relationship has been observed in dwarf galaxia fish (*Galaxiella pusilla*) (Box 7.2), mice, and several Australian mammals, and is expected to apply generally. Thus, high genetic differentiation often identifies populations that have lost genetic diversity and would benefit from gene flow, rather than populations that should be managed separately.

We seek a paradigm shift whereby evidence of genetic differentiation among populations triggers questions of whether any population segments are suffering genetic problems, and whether they can be rescued by gene flow, rather than routinely recommending that segments be managed separately.

Box 7.2 High genetic differentiation among dwarf galaxia fish populations is causally associated with low genetic diversity within populations

(after Frankham et al. 2017, Box 10.2, based on Coleman et al. 2013)

Threatened dwarf galaxia fish were sampled from 51 populations (black dots) in different rivers in southeastern Australia and genotyped for microsatellites. Fish populations with the highest genetic differentiation with other populations (F_{ST}) exhibited the lowest microsatellite heterozygosity (H_e) within populations and were the ones most in need of gene flow. This was true in both the western and eastern populations (left and right figures below).

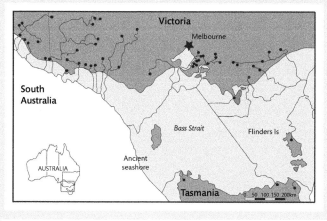

Dwarf galaxia fish (Australia)

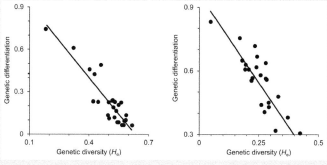

What are our objectives?

Objectives

We seek to determine if population fragments are suffering genetic erosion and whether any would benefit from gene flow.

What questions should we ask, having established the presence of multiple populations?

The decision tree in Fig. 7.1 provides a perspective on where we are heading in the genetic management phase of the book. The first step is to determine how many genetically distinct populations exist and where they are located (Chapter 4). Questions 2 through 5 in the decision tree are covered in this chapter.

Fig. 7.1 Decision tree for genetic management of fragmented populations, designed to guide us through the genetic management chapters (after Frankham et al. 2017, Fig. 10.1).

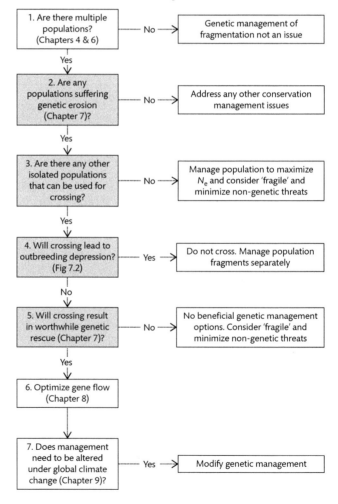

Box 7.3 illustrates the considerations and evidence used in evaluating the case for genetic rescue of the endangered Florida panther.

Box 7.3 On what basis was the decision made to attempt genetic rescue of the Florida panther?

(after Frankham et al. 2017, Box 11.1, based on Roelke et al. 1993; Hedrick 1995: Culver et al. 2000, 2008)

By the early 1990s, the endangered Florida panther was restricted to a small relict population of ~ 20–25 individuals in southern Florida. Prior to European settlement, they ranged across the entire southeastern USA, and other named sub-species were spread throughout North and South America.

The following are the considerations that went into the decision to attempt genetic rescue of this population.

Was the Florida panther suffering genetic erosion?

The Florida panther population had low levels of genetic diversity compared to earlier museum specimens, other panther populations, and felids generally, so it was inbred.

The panthers showed signs of inbreeding depression, including kinked tails (photograph), cardiac defects, a high prevalence of infectious disease, and very poor semen quality, compared to other panthers and felids. Further, about half of the males had at least one undescended testis and the incidence had increased over time.

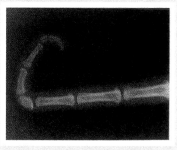

Florida panther (USA)

(After Roelke et al. 1993)

Were there other populations that could be used to rescue the Florida panthers?

Yes, this species is widely distributed throughout the USA. In particular, the Florida and Texas populations had previously been connected.

What was the risk of outbreeding depression in a cross between Florida and Texas animals?

Because Florida and Texas animals were designated as distinct sub-species at that time, outbreeding depression was considered a potential problem. However, the

Continued ➤

---

main distinguishing feature of the Florida panther, its kinked tail, was likely a manifestation of inbreeding depression. A subsequent taxonomic revision included Florida panthers with all other North American populations in the *Puma concolor cougar* sub-species.

While the habitats in Florida and Texas are somewhat different, the species is wide-ranging and inhabits diverse environments throughout North and South America. Concerns about possible chromosomal divergence were not mentioned, but all felid species have similar karyotypes, so fixed chromosomal differences are unlikely.

What was the final decision?

Following extensive consultations, population viability analyses (PVAs), and genetic modeling studies, the Florida population was augmented with eight wild-caught Texas puma females. Box 5.1 details the subsequent beneficial consequences of this restoration of gene flow on fitness and genetic diversity.

Before addressing our management objectives, we provide existing guidelines for genetic rescue attempts.

Guidelines for genetic rescues

We endorse the guidelines in Table 7.1 because they reflect advances in predicting the risk of outbreeding depression, meta-analyses on genetic rescue, and other relevant advances. These replace earlier and more restrictive guidelines for genetic rescue.

Table 7.1 Guidelines for management of genetic rescues (after Frankham 2015).

When should we contemplate genetic rescue and how should we do it?
When should we contemplate genetic rescue?   1. When there is a (recipient) population that is inbred and/or has low genetic diversity, especially when it is known or suspected to be suffering from inbreeding depression for fitness.   2. When there is another genetically isolated (donor) population(s) of the same species to which it can be crossed to reverse inbreeding and restore genetic diversity.   3. When the risk of outbreeding depression in crosses between the donor and recipient populations is expected to be low through to the F_3 generation, as determined for example by using our decision tree (Fig. 7.2).

4. When the potential benefits of genetic rescue are sufficiently large to justify the risks and costs of translocation and any risk of outbreeding depression. The more inbred the population, the larger the benefits. The benefits also depend on:
   a. The mating system in the species (outbreeders > selfers)
   b. The intended environment (wild > captive)
   c. Inbreeding level in immigrants (outbred > inbred)
   d. Demographic history and genetic diversity of the original population from which the recipient and donor populations were derived (numerically large [N_e] population with high genetic diversity > numerically small with low genetic diversity).

How should we do it?

1. How many immigrants should be used? Any are better than none when the risk of outbreeding depression is low. At the upper end there is a risk of genetically swamping the recipient population, so immigrant alleles should generally be ≤ 50% of the crossed population (see Chapter 8).
2. Will more than one augmentation of gene flow be required? The need for additional rounds of crossing will depend upon the proportion of immigrants (low > high), their inbreeding level (high > low), the N_e in the crossed population (low > high), and the number of generations since the previous augmentations (many > few).
3. Should the results of the program be monitored? Yes.

We next turn to answering questions 2 through 5 of Fig. 7.1.

Question 2: Are any populations suffering genetic erosion?

To determine whether a population is likely to be suffering from problematic levels of genetic erosion in fitness or ability to evolve, we need evidence for one or more of the following:

- lower fitness than in a related non-inbred population
- lower molecular genetic diversity than in a related non-inbred population
- elevated pedigree inbreeding coefficient
- mean $F > 10\%$ estimated from population size and inferred N_e and number of generations of isolation (see Box 7.4 later).

If an isolated outbreeding population has suffered at least a 10% loss of fitness, a known or inferred 10% inbreeding coefficient, or 10% loss of genetic diversity, we conclude that it is likely suffering from genetic erosion. The thresholds of at least 10% follow those for tolerable losses of genetic diversity over 100 years used for many captive populations of threatened species (Frankham et al. 2010), and those for inbreeding depression in the wild under the Frankham et al. (2014) guidelines. An inbreeding level of 10% may seem like a low threshold for intervention, but it leads to a reduction in total fitness of 45% compared to an outbred population, assuming an average of six diploid lethal equivalents per genome (Chapter 3).

In many cases, there may not be direct information on inbreeding depression and genetic erosion, and those will have to be assessed from more indirect information. There are four main approaches for obtaining estimates of inbreeding when there are no pedigrees. First, kinship and individual inbreeding coefficients can be estimated from genotypes for multiple SNPs or microsatellites based on identity by descent of alleles in individuals within the population, using software such as COANCESTRY. This method has been used to estimate kinship in the founders of the California condor (*Gymnogyps californianus*) captive population, based on SNPs. Improved genomic methods based on the length of chromosomal sections where all genetic markers are homozygous (termed runs of homozygosity) can be used to estimate *F* back to remote common ancestors and have been validated in birds and mammals. These are recommended if possible, but require an assembled genome and data on many thousands of SNPs in each individual (Kardos et al. 2018; Nietlisbach et al. 2018). These and other genomic methods will typically be more accurate than those from the following methods.

Second, population mean inbreeding coefficient can be estimated from heterozygosities for "neutral" markers such as SNPs or microsatellites, in the target populations and in an outbred population, as illustrated in Example 7.1.

Black-footed rock-wallaby (Australia)

Example 7.1 Estimating the inbreeding coefficient in the Barrow Island rock-wallaby population from microsatellite data

(Frankham et al. 2017, Example 11.1)

The Barrow Island population of black-footed rock-wallabies has an average microsatellite heterozygosity of 0.05 (H_{Inbred}), while two mainland populations have an average heterozygosity of 0.59 (H_{Outbred}) (Eldridge et al. 1999). Consequently, the estimated inbreeding coefficient (*F*) for the Barrow Island population is:

$$F = 1 - \frac{H_{\text{Inbred}}}{H_{\text{Outbred}}} = 1 - \frac{0.05}{0.59} = 0.915$$

Thus, the Island population is highly inbred. Further, the frequency of lactating females in Island animals is only 56%, compared to 92% in mainland ones, suggestive of inbreeding depression.

Third, population mean *F* can be estimated from the population size and generations of isolation, as illustrated in Box 7.4. While the N_e is generally unknown, as a first approximation it can be obtained from the mean adult population size and N_e/N ratios from related species (Chapter 2 and Appendix 2).

Box 7.4 Estimating the mean inbreeding coefficient for isolated populations of greater bamboo lemurs from population size and generations of isolation

(after Frankham et al. 2017, Box 11.2; lemur information from Tony King, pers. comm.)

Small isolated populations of critically endangered greater bamboo lemurs (*Prolemur simus*), mostly consisting of ~ 5–30 individuals, occur in deforested lowland agricultural landscapes in Madagascar.

A completely isolated population of 30 individuals would likely have an effective size (N_e) of ~ 3–10 (Chapter 2). The "isolated" sites have probably been isolated for less than 100 years. With a generation time of 14 years (the mid-point of female breeding duration from ~ 3 to ~ 25 years), there have been ~ 7 (100/14) generations of isolation (t).

Given seven generations of isolation at an N_e of 10, the inbreeding coefficient (F) in our target population will be:

$$F = 1 - (1 - \frac{1}{2N_e})^t = 1 - (1 - \frac{1}{2 \times 10})^7 \sim 0.3$$

If N_e is only 3, then $F \sim 0.72$. These two estimates should bracket the true value of F (providing there is no gene flow), and both indicate that augmentation of gene flow should be considered.

Greater bamboo lemur (Madagascar)

Fourth, computer modeling (e.g. using VORTEX [see Appendix 4]) can be used to predict the decline in heterozygosity and the inbreeding coefficient from trajectories in adult population sizes. For example, the simulated decline of heterozygosity in the introduced banteng (*Bos javanicus*) population in northern Australia was similar to that determined from microsatellite analyses.

Question 3: Is there another population to which it can be crossed?

Candidate populations for use in genetic rescue will typically belong to the same species and have been genetically isolated from the recipient population for a substantial number of generations, with the amount of genetic differentiation determined by the combinations of effective population sizes and the number of generations. Outbred populations are better sources than inbred ones, and the best population may be a different sub-species.

Total genetic isolation is not necessarily required, but one of the populations should not be a recent derivative of the other. However, if there is no option, there may be some benefits in gene flow into a bottlenecked population from its larger source population, as has been done with the Mauna Kea silversword (Robichaux et al. 1997; Friar et al. 2000).

If there is no population to which the inbred population can be crossed, its size should be increased as much as possible, and management should consist of non-genetic procedures designed to minimize the risk of extinction (reduction of threats, fire management, disease sanitation, captive propagation, increase population size, etc.), as applied in the Wollemi pine (*Wollemia nobilis*) and black-footed ferrets (*Mustela nigripes*).

If there is another population to which the inbred population could be crossed, we should next evaluate the risk of outbreeding depression in the cross.

Question 4: Will gene flow result in outbreeding depression?

We developed a decision tree with five questions to predict the risk of outbreeding depression (Fig. 7.2), and it has correctly identified the risk of outbreeding depression resulting from between population crosses for almost all cases examined (Chapter 5).

Fig. 7.2 Risk of outbreeding depression (OD) decision tree for determining whether the crossing of populations is likely to be harmful (after Frankham et al. 2011).

Populations with reliable taxonomy, no fixed chromosomal differences, gene flow within the last 500 years, and inhabiting similar environments have low risks of outbreeding depression following crossing

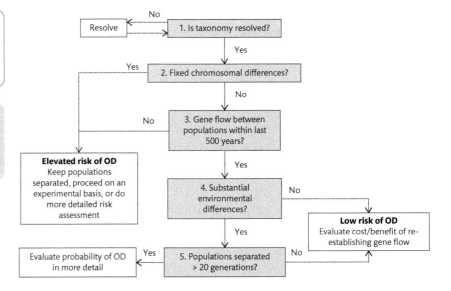

For example, isolated diploid populations of the threatened button wrinklewort daisy from the Canberra region in southeastern Australia have the attributes specified in the margin box above and exhibited improved fitness when populations were crossed, as do sub-species of old-field mice (Table 5.5).

Button wrinklewort daisy (Australia)

Conversely, populations with high risk of outbreeding depression when crossed have fixed chromosomal differences, or have not exhibited gene flow in the last 500 years, or are adapted to different environments, or have a combination of these attributes (Table 5.5), and we generally recommend that they be managed separately or have their risk evaluated in more detail.

We recommend that populations with > 20 generations of adaptation to different environments (but no other risk factors) be subject to a detailed assessment of risk, based upon their genetic diversities, effective population sizes, how different their environments were, and the number of generations since they were isolated.

The risk of outbreeding depression from between population crosses increases with generations of isolation. We used a 500-year time span to simplify this, as this captures much of the European human-associated habitat loss, fragmentation, and degradation that is our primary concern (see Preface).

The question about gene flow within the last 500 years is designed to avoid outbreeding depression from crosses between morphologically similar but reproductively isolated cryptic species

This 500-year recommendation errs on the side of caution, because it is often possible to cross populations that have been separated longer in similar environments, with no adverse effects. For example, crossing of mountain pygmy possum (*Burramys parvus*) populations isolated for at least 20,000 years has resulted in improved fitness (Weeks et al. 2017). Thus, we would not rule out a genetic rescue based solely on the time two populations have been separated, but such cases should be subject to scrutiny by experts.

Historical or molecular genetic data can be used to estimate the duration of isolation between populations

Historical information may indicate when populations became isolated. For example, habitat fragmentation due to the impacts of European settlement in Australia began in 1788 (when the first European settlers arrived). This may be distinguishable from fragmentation caused from aboriginal activities beginning ~ 60,000 years ago by using information in diaries of early explorers and traditional knowledge.

Alternatively, divergence time can be estimated using modeling based on molecular genetic data. For example, divergence times between brown trout (*Salmo trutta*) populations above and below old dams in a river system in Denmark were estimated using two different modeling approaches applied to data from 40 microsatellites. Both models suggested divergence times ~ 600–800 years ago, concordant with the historical record of dam building.

Brown trout (Europe)

Similar versus different environments can be defined by expert opinion or by ecological niche modeling

Expert scientific opinion can be used to assess whether environments are similar or different and would lead to differential local adaptation, as it is widely used in planning for translocations and reintroductions. Expert opinion should be elicited using recent guidelines that minimize biases in human perception (Kahneman 2011; Martin et al. 2012). Determinations should rely on the range of variation of key features of the two environments, such as the non-overlap of environmental characteristics to which a species is adapted or sensitive, and whether the species is a narrow specialist (such as some butterflies and plants) or a generalist (such as some wide-ranging birds and large carnivores).

Ecological niche modeling (also termed multidimensional environmental scaling) is used to associate species or populations with habitat envelopes (such as precipitation, elevation, mean annual temperature, and vegetation), using software such as BIOCLIM and GARP.

How should clines be considered?

For populations that previously exhibited clines and are now fragmented, we recommend gene flow between nearby populations inhabiting similar environments, in preference to distant ones that may inhabit different environments

In general, there is an increased risk of outbreeding depression when populations from increasingly distant geographic locations are crossed, especially when there are clines and adaptive differences. Consequently, if species that previously exhibited continuous clines now exist in isolated fragments that need to be augmented, gene flow should come from nearby populations rather than distant ones, as was done for red-cockaded woodpeckers (Chapter 4 frontispiece and Box 4.1).

We do not preclude gene flow between two populations that share a chromosomal arrangement, but one is polymorphic and the other monomorphic

Some species have some populations that are polymorphic for chromosomal arrangement, while others share only one of the arrangements, as for example occurs for centric fusions in some common shrew (*Sorex araneus*) populations in Europe, and for an inversion in some populations of Frémont's western rosinweed plant (*Calycadenia ciliosa*) in the western USA. Crossing a polymorphic population to a monomorphic one should not lead to outbreeding depression and may be beneficial. Further, crossing populations with different numbers of accessory chromosomes (e.g. supernumerary or non-pairing B chromosomes) should pose little risk of significant outbreeding depression (Lanzas et al. 2018).

Diploid and polyploid populations do not represent a polymorphism and should not be crossed.

If information required to answer a question is missing, we recommend passing to the next question and completing a preliminary risk assessment

If the information required to answer a question in the decision tree is missing, as will often occur for chromosomes, we recommend that managers pass to the next question and complete a preliminary assessment. If this yields a low probability of outbreeding depression, it is advisable to obtain the missing information or to proceed to use experimentally instituted gene flow. Obtaining chromosomal information is particularly important in groups with varying ploidies (e.g. plants) or high rates of chromosomal evolution (e.g. primates and rock-wallabies).

If the risk of outbreeding depression is low, the next step is to determine whether the benefits of gene flow will be sufficient to justify a genetic rescue attempt.

Question 5: Will gene flow result in worthwhile genetic rescue?

The potential benefits of gene flow on fitness depend upon:

- whether the intended environment is more or less stressful (more stressful > less)
- the mating system (naturally outbreeding > inbreeding)
- inbreeding depression in the target population (high > low)
- whether the donor population is inbred or outbred (outbred > inbred)
- the extent to which gene flow reduces inbreeding (more > less).

Thus, the benefits of genetic rescue are likely to be large if the target population is in a wild environment, does not naturally inbreed, but is highly inbred and probably suffering from inbreeding depression, and it is possible to introduce enough individuals or gametes from an outbred donor population to substantially reduce its inbreeding levels. As we saw in Chapter 5, a meta-analysis found that gene flow into inbred populations was beneficial in over 90% of cases and that wild populations showed a median improvement of nearly 150% in composite fitness. In many cases, a qualitative assessment of the probable benefits may be enough to support a recommendation for a genetic rescue or a feasibility study for a genetic rescue. Those who consider this sufficient can skip the next two sections, as they contain quantitative predictions based on use of equations. However, for those who want **project-specific predictions of the benefits**, we provide methods in the following two sections, so you can complete quantitative cost–benefit analyses.

Predicting the magnitude of genetic rescue effects for naturally outbreeding species requires two steps. First, we need to determine the reduction in inbreeding (or, equivalently, the increase in genetic diversity) that will be obtained from the augmented gene flow. Second, we need to estimate how much fitness benefit will be obtained from that reduced inbreeding.

Determining the change in inbreeding due to different gene flow regimes

Example 7.2 provides a method for estimating the inbreeding coefficient that results from gene flow into an inbred population. Two examples are given, the former when gene flow is from one other population, and the latter where a population is formed by crossing several different populations with diverse inbreeding levels in possibly dissimilar proportions (relevant to later considerations). The inbreeding coefficient for the crossed population (F_{pooled}) is that attained after random mating is established in the F_2 generation, at which point inbreeding will be relatively stable in large random-mating populations.

Example 7.2 Computation of inbreeding coefficients for populations formed by pooling populations with different F values

(Frankham et al. 2017, Example 11.3)

For the cross of a completely inbred population ($F = 1$) to an outbred one ($F = 0$), the inbreeding coefficient in the crossed population is computed as follows, where f_i is the proportional contribution from the two populations to the pooled one, and n is the number of contributing populations:

$$F_{pooled} = \sum_{i=1,}^{n} f_i^2 F = (0.5^2 \times 1) = 0.25$$

Thus, the crossing of the populations results in a 75% reduction in inbreeding, while crossing of two independent completely inbred populations results in a 50% reduction in inbreeding (Table 5.1).

If we pool three populations with F values of 0.1, 0.3, and 0.5 in the proportions of 0.5, 0.4, and 0.1, the F of the pooled populations after random mating equilibrium is established is computed as follows:

$$F_{pooled} = \sum_{i=1,}^{n} f_i^2 F = (0.5^2 \times 0.1) + (0.4^2 \times 0.3) + \left(0.1^2 \times 0.5\right) = 0.078$$

Thus, the pooled population has an inbreeding coefficient of 7.8% in the F_2.

We next use the changes in inbreeding due to gene flow to predict the changes in fitness.

Determining the fitness benefit obtained from a given decrease in inbreeding

When a more specific prediction of the probable benefits of genetic rescue for a target population is desired, it can be obtained either by assuming that the target population contains a specific number of lethal equivalents (Example 7.3) or, if the inbreeding depression in the target population is known, from the expected reduction in inbreeding resulting from the gene flow (Example 7.4).

Both methods rely on the expectation that the natural logarithm (ln) of relative fitness (compared to an outbred population with $F = 0$) will typically decline linearly with the amount of inbreeding, as it is widely assumed to do for survival (Chapter 3; Ralls et al. 1988; O'Grady et al. 2006; Frankham et al. 2014). The reduction in inbreeding and increase in relative fitness due to gene flow is expected to involve a reversal of the inbreeding depression trajectory (Frankham et al. 2017). As discussed in Chapters 3 and 5, it is important to estimate relative fitness in the environment that will be experienced by the population (or as close to it as feasible).

Example 7.3 Predicting the magnitude of genetic rescue effects using lethal equivalents

If the fitness assessment is carried out in a wild environment, lethal equivalents (B) value will typically be on average ~ 6 (Table 3.1 and related text in Chapter 3). For a reduction in inbreeding coefficient from 0.4 in the inbred (F_I) to 0.2 in the cross (F_X) and $B = 6$, we predict a genetic rescue ΔGR of:

$$\Delta GR = e^{(F_I - F_X)B} - 1 = e^{0.2 \times 6} - 1 = 2.32$$

Thus, we predict 232% improvement in total fitness in the F_3 and subsequent generations compared to the inbred parent.

The improvement in total fitness is expected to be less if the population contains fewer lethal equivalents, as for example due to a history of small population size (Box 8.1) or if the environment is benign.

The second method is based on the expectation that the natural logarithm of relative fitness will increase by the same proportion that the inbreeding coefficient is reduced by gene flow (Chapters 3 and 5), as illustrated in Example 7.4.

Example 7.4 Predicting the extent of genetic rescue for fitness when inbreeding depression is known

In this example, we ask by how much relative fitness will increase if an inbred population suffering inbreeding depression of 60% (relative fitness 0.4) is crossed to an unrelated equally inbred population, which halves the inbreeding coefficient (Table 5.1). If the relative fitness of the crossed population in the F_3 is designated W_X and that of the inbred parent as W_I (0.4) and inbreeding is halved ($F_X/F_I = \frac{1}{2}$), then:

$$\ln W_X = \left(\ln W_I\right)\left(\frac{F_X}{F_I}\right) = \left(\ln 0.4\right)\left(\frac{1}{2}\right) = \frac{-0.916}{2} = -0.458$$

Thus, $W_X = e^{-0.458} = 0.63$.

Consequently, the relative fitness is predicted to rise from 0.4 to 0.63 (58%) as a result of this genetic rescue regime. Similarly, reducing inbreeding by ¾ by crossing to an outbred donor population (Example 7.2) is predicted to improve relative fitness by 99%.

Improvements in the ability to evolve due to crossing are predicted to be proportional to the improvement in heterozygosity

The benefits of gene flow exist not only in the increased fitness due to reduced inbreeding in the subsequent few generations, but also in the increased ability of the population to evolve adaptively in response to stresses and changing environments. The improvement in ability to evolve increases with the increase in genetic diversity (Fig. 5.2 and associated text). Thus, the proportionate increase in ability to evolve is predicted to be (heterozygosity$_{F2 \text{ cross}}$/heterozygosity$_{Recipient}$) – 1 (Fig. 5.2). For example, crossing the small inbred Mt. Jasmin population of the jellyfish tree to the Bernica population in the Seychelles was expected to increase the microsatellite heterozygosity from 0.26 to 0.59, yielding a 127% expected increase in ability to evolve. This can also be computed from the inbreeding coefficients as ($F_{Recipient}$ – F_{F2})/(1 – $F_{Recipient}$).

Where beneficial and harmful effects of gene flow are both expected, a risk–benefit analysis is recommended

In some crosses, beneficial and harmful impacts occur simultaneously. For example, both impacts were found in crosses of beach and old-field *Peromyscus* sub-species, of different populations of the common *Primula vulgaris*, and in long-distance crosses of partridge peas (Barmentlo et al. 2018). Since we are not able to make quantitative estimates of the risk and magnitude of outbreeding depression, we can only evaluate relative risks.

If the risk of outbreeding depression is modest, but the expected benefits of gene flow are large, then genetic rescue attempts are desirable. For example, crossing very divergent populations of Trinidadian guppies (*Poecilia reticulata*) produced a beneficial effect on fitness (Kronenberger et al. 2017). Conversely, if there is a significant risk of outbreeding depression and only small expected benefits, as is more likely in a selfing species, initiating gene flow is unlikely to be warranted.

In Chapter 8 we address choice of the best population(s) for genetic rescue, and the management of rates of gene flow, based on different amounts and types of information.

Summary

1. The current practice when populations are genetically differentiated is usually to recommend separate genetic management, **but this is often ill-advised and a paradigm shift away from this is needed**. Instead, the advisability of augmenting gene flow should be addressed, as follows.
2. Small isolated populations that are highly divergent from other populations are often inbred and would benefit from genetic rescue rather than separate management.
3. We should first ask whether the target population is inbred or has suffered loss of genetic diversity, and is it suffering from known or suspected inbreeding depression.
4. If a population is suffering genetic erosion, we should ask if there are any donor populations within the species to which it could be crossed.
5. Next we ask what the risk of outbreeding depression is in crosses between the two populations. This will be low for populations with the same karyotype, that are adapted to similar environments, and that were isolated within the last 500 years.

6. If a cross has a low risk of outbreeding depression, the potential fitness benefits depend upon the extent of inbreeding depression in the target population (large > small), the size of the difference in inbreeding between the inbred parent population and the cross in the F_3 generation (large > small), the environment (stressful > benign), inbreeding level in the donor population (outbred > inbred), and the mating system (outbreeders > inbreeders).

7. The improvement in ability to evolve after gene flow can be predicted from the ratio of heterozygosity in the F_2 cross to that of the inbred parents.

FURTHER READING

Frankham (2015) Meta-analysis that (a) evaluated the effectiveness of a screen against outbreeding depression, (b) documented the magnitude of genetic rescue effects, and (c) provided updated guidelines for genetic rescue.

Frankham et al. (2011) Reviews causes of outbreeding depression, presents a decision tree for assessing its risk of occurrence, and provides evidence that it works.

Frankham et al. (2017) *Genetic Management of Fragmented Animal and Plant Populations*: Chapters 10 and 11 have more detailed treatment of topics in this chapter.

Wang (2011), Keller et al. (2011), and Knief et al. (2015) Describe methods for estimating levels of inbreeding from SNP data that are applicable to species in the wild.

SOFTWARE

BIOCLIM: the first species distribution modeling package (Booth et al. 2014). https://fennerschool.anu.edu.au/research/products/anuclim

COANCESTRY: software to estimate kinship and inbreeding coefficients from multilocus genotype data (Wang 2011). http://www.zsl.org/science/software/coancestry

GARP: desktop species distribution modeling software (Stockwell 1999). https://www.gbif.org/tool/81288/desktopgarp/

PLINK: contains software for estimating inbreeding coefficients from runs of homozygosity (Purcell et al. 2007). https://www.cog-genomics.org/plink2/

VORTEX: population viability analysis software that tracks the predicted heterozygosity and inbreeding coefficients of individuals (Lacy & Pollak 2014). https://scti.tools/vortex

Managing gene flow among isolated population fragments

When the decision is made to initiate gene flow into an isolated population, managers must decide when to start, from where to take the individuals or gametes, how many, which individuals, how often, when to cease, etc. Even without detailed genetic data, sound management strategies for augmenting gene flow can be developed by considering conservation genetics theory or using computer simulations. Moving some individuals into isolated inbred population fragments is better than moving none. With more detailed genetic information, more precise genetic management of fragmented populations can be achieved, leading to improved genetic outcomes. Gene flow management will be most effective if done using mean kinship (estimated from modeling, genetic markers, or pedigrees), and moving individuals from the fragment with the lowest mean kinships into the target fragment(s). Further improvements can be made using individual kinships to choose the best individuals to move. Populations should then be monitored to confirm that movement of individuals has enhanced genetic diversity and fitness.

TERMS

Corridor, genetic swamping, kinship matrix, panmictic

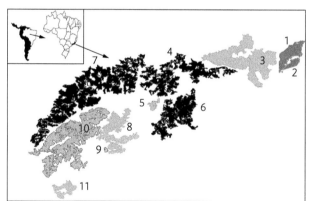

The golden lion tamarin's fragmented population in Brazil. This species' extensive conservation program includes translocations, building corridors, captive breeding, reintroduction, and habitat restoration. Fragments shaded black (4, 6, and 7) were the original wild populations. The lightly shaded fragments (3, 5, 8, 9, 10, and 11) were founded and managed with gene flow from the captive population. The mid-shaded fragments 1 and 2 in the upper right were formed by translocating individuals from several extremely small vulnerable wild fragments (Frankham et al. 2017, p. 266). (Image provided by the Associação Mico-Leão-Dourado.)

A Practical Guide for Genetic Management of Fragmented Animal and Plant Populations. R. Frankham, J. D. Ballou, K. Ralls, M. D. B. Eldridge, M. R. Dudash, C. B. Fenster, R. C. Lacy & P. Sunnucks. Oxford University Press (2019). © R. Frankham, J. D. Ballou, K. Ralls, M. D. B. Eldridge, M. R. Dudash, C. B. Fenster, R. C. Lacy & P. Sunnucks 2019. DOI: 10.1093/oso/9780198783411.001.0001

Genetic management of gene flow among population fragments is needed to optimize genetic benefits in a cost-effective manner

Gene flow can range from a single immigrant to so many that the recipient population is essentially replaced. Appropriate levels of gene flow are needed to rescue fitness and genetic diversity, to avoid genetic replacement of the original population, and to minimize the cost and frequency of augmentation. The number of initial immigrants affects how soon more gene flow is needed. Should we have large gene flow now, lower levels every generation, or something in between? How do we choose the best population(s) for genetic supplementation? These proposals have to be integrated with non-genetic issues, such as the organism's life history, behavior, and which sex and age should be used to augment gene flow.

We have previously determined that a population needs genetic augmentation and have identified candidate donor populations with a low risk of causing outbreeding depression. This chapter describes methods to choose the best population(s) for genetic rescue and to manage gene flow between populations, depending on how much genetic and demographic information is available. Box 8.1 illustrates how computer modeling can aid genetic management of Allegheny woodrats (*Neotoma magister*), as a prelude to detailed treatment of the issues later in this chapter.

Box 8.1 Computer modeling of genetic management options for fragmented populations of Allegheny woodrat

(after Frankham et al. 2017, Example 12.1, based on information from T.J. Smyser, pers. comm.)

Allegheny woodrats in southern Indiana, USA are restricted to eight small habitat fragments, with few close enough to allow natural dispersal. Computer modeling using VORTEX showed that if the fragments were interconnected (panmictic), the single large population would quickly grow to carrying capacity (top line in figure), would be demographically resilient, and would retain 91% of its initial heterozygosity for 100 years (mean $F = 0.09$ [table]). However, without intervention the woodrats become highly inbred (mean final $F = 0.37$), and assuming four lethal equivalents (presuming some purging due to past inbreeding), most subpopulations crash to extinction and the metapopulation declines markedly in size (bottom line in figure). Local inbreeding depression caused its decline, because simulations without inbreeding depression (middle line) had demographic performance similar to that of the panmictic metapopulation (despite F rising to $= 0.34$).

Allegheny woodrat (USA)

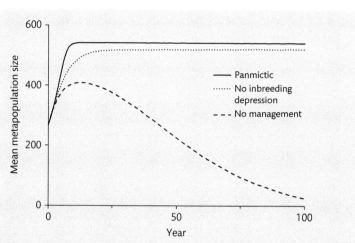

Moving one to eight random individuals per generation between fragments (Nm = 1–8) reduced inbreeding, decreased sub-population extinctions, and improved population growth (next graph and table).

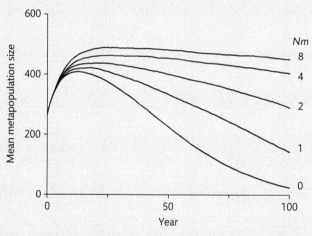

Managed translocations	Mean probability of sub-population extinction	Final metapopulation N	Mean inbreeding (F)
$Nm = 0$	0.81	26	0.37
$Nm = 1$	0.18	144	0.24
$Nm = 2$	0.02	288	0.18
$Nm = 4$	0.00	404	0.14
$Nm = 8$	0.00	450	0.11
Panmixia	0.00	540	0.09

Gene flow can be established by connecting habitat fragments or moving genetic material between them

A long-term solution to inadequate gene flow among fragmented populations is to improve the habitat matrix between them or establish habitat corridors. For example, habitat corridors have been established between isolated habitat fragments for the golden lion tamarin in Brazil. Modifying fences, removing dams or roads, or reducing grazing or pesticide use may also improve gene flow. Wildlife underpasses or overpasses are being used to re-connect populations of terrestrial animals, as in Banff National Park in Canada (Fig. 8.1), and canopy bridges, glider poles, and vegetated medians in Australia.

If restoration of gene flow through habitat management is not practical, human intervention is required to move individuals or gametes. Both approaches can be combined, beginning with translocations for initial genetic rescue, and initiating habitat management to improve connectivity to sustain genetic rescue.

Fig. 8.1 Highway overpass built as a wildlife corridor over the Trans-Canada Highway in Banff National Park (2012, http://conservationcorridor.org/2012/10/banff-national-park/). (Photo courtesy of Adam Ford, ARC Solutions http://arc-solutions.org.)

In the remainder of this chapter, we assume that movement of individuals or gametes is necessary to augment gene flow, and we offer guidance for implementing this option. The first step is to define the objectives of managing gene flow.

Objectives

Our objectives are to maximize genetic diversity and minimize inbreeding by establishing gene flow in a cost-effective manner.

What questions should we ask?

To achieve these objectives, we need to answer the following questions:

- From which donor population(s) should gene flow be established?
- Should augmentation be from one or several populations?
- Should we use inbred or outbred donor populations?
- How many individuals or gametes to move?
- Which sex and age to move?
- Which individuals to move?
- When should genetic augmentation begin?
- When to stop?
- Will genetic rescue be required again in the future?

Our ability to answer these questions depends on the information available on the managed population(s). We examine situations where the amount of information varies from little to detailed (Fig. 8.2), encompassing programs with limited funds and basic information, to well-funded programs with access to up-to-date genomic techniques where more precise management of gene flow is possible.

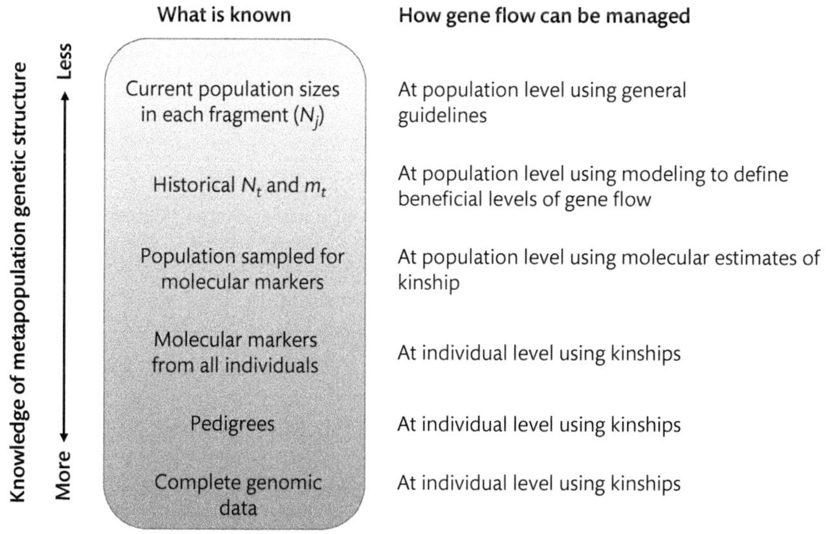

Knowledge of metapopulation genetic structure (Less ↑ / More ↓)	What is known	How gene flow can be managed
	Current population sizes in each fragment (N_j)	At population level using general guidelines
	Historical N_t and m_t	At population level using modeling to define beneficial levels of gene flow
	Population sampled for molecular markers	At population level using molecular estimates of kinship
	Molecular markers from all individuals	At individual level using kinships
	Pedigrees	At individual level using kinships
	Complete genomic data	At individual level using kinships

Fig. 8.2 Types of gene flow management possible with different levels of information on current (N_j) or historical population size (N_t), migration rates (m_t), population structure, and kinships (after Frankam et al. 2017, Fig. 12.2).

The following section illustrates options for genetic management of fragmented populations when only basic information is available (Fig. 8.2).

Management of gene flow with minimal information

With minimal genetic information, we can rely only on general principles to choose the donor population and manage gene flow (Chapters 3–5 and 7).

From which donor population(s) should gene flow be established?

Augmenting from any isolated population that has passed the screening in Chapter 7 is better than not augmenting. Four issues contribute to the choice of donor population:

- Donor and recipient populations should be substantially differentiated from each other
- Large donor populations should be superior to small ones, assuming they have higher genetic diversity
- Outbred donors should be better than inbred ones
- Several donor populations should be better than a single one, assuming that jointly they will contribute more new genetic diversity.

Closely related populations, such as those with source–sink relationships, or those in a metapopulation where one population has been recently derived from the other, should be avoided if possible because gene flow would provide little benefit. Larger populations should contribute more genetic diversity to the recipient population. Outbred or inbred donor populations can be used, but outbred ones have greater genetic benefits, and their use requires less frequent augmentation. Gene flow from several donor populations should add more genetic diversity and reduce inbreeding further, and lead to benefits that persist better across generations.

There may be other logistical considerations that modify the choice of source populations, especially in large species, e.g. geographical distance, political boundaries, cost of translocating individuals, cuttings, or gametes, and flowering synchronization in plants.

Which individuals?

In the absence of genetic information, individuals should be chosen to minimize the chance of selecting close relatives, and should be sampled across the distribution of the population to capture maximum genetic diversity

The approach described in the margin box is likely to achieve considerable benefit, but likely less than obtainable from a more refined analysis. In species with two sexes, either sex (or both) can be used in genetic rescue, but one sex may be preferred for genetic, ecological, demographic, or behavioral reasons. Young animals have greater lifetime reproductive potential, are often natural dispersers, and may be better accepted by resident animals, so are often the best to translocate. In plants, pollen, seeds, cuttings, or transplants may be used to produce gene flow, but transplants may have a better chance of establishing and reproducing. Guidelines that consider other aspects of reintroductions and translocations are provided by IUCN/SSC.

How many?

When the recipient population is inbred, any immigrants are better than none

Only a modest number of successfully reproducing immigrants each generation is needed to meaningfully improve fitness and genetic diversity, even a single individual (Box 8.1). When a small population is inbred, failing to increase gene flow is a risky strategy likely to lead to eventual extinction.

The optimal number of immigrants depends on whether the goal is to avoid damaging levels of inbreeding from accumulating, provide genetic variation for adaptation to changing environments, reversing substantial existing inbreeding, or reducing inbreeding to acceptable levels within a single generation. We discuss each of these below.

At least five contributing immigrants per generation are recommended to prevent damaging inbreeding

Even though a single immigrant that breeds can reduce inbreeding and improve fitness, one per generation is insufficient to prevent substantial inbreeding from accumulating over time, because it results in an inbreeding coefficient of at least 0.2 in the long term.

To keep inbreeding coefficients below 10% in real populations (Chapter 7), **we recommend that typically five genetically contributing immigrants be introduced per generation.** This will be at least 40 actual immigrants based on an average N_e/N ratio of 0.125. The Allegheny woodrat requires just over eight immigrants per generation into each population fragment to keep the overall inbreeding level below 10%, but because the model assumes that immigrants have the same survival and reproductive rates as the residents, not all of them would breed (Box 8.1 table).

Gene flow necessary to maintain adequate variation for adaptive evolution is at least as large as that needed to avoid damaging inbreeding

If we assume that more than a 10% decline in the rate of adaptation could place a population at risk, then the gene flow required to sustain adaptive potential will be approximately the same as necessary to avoid damaging inbreeding. However, under global climate change many species will need to adapt more rapidly than previously, so the numbers of immigrants may need to be even greater (see Chapter 9).

With a single generation of gene flow, many unrelated immigrants are typically needed to reduce inbreeding

To estimate the number of unrelated immigrants needed to reduce inbreeding with a single generation of augmentation, we first need to estimate the current level of inbreeding from the population size and the number of generations of isolation (Example 3.1). When the goal for genetic rescue is to reduce the inbreeding coefficient in the pooled population to 0.10, the relationship between immigrants needed and inbreeding in the fragment is shown in Fig. 8.3. For example, it takes 15 non-inbred immigrants to reduce F from 0.2 to 0.1 in a population fragment with 50 breeding adults (Example 8.1).

Fig. 8.3 The proportion of non-inbred, unrelated immigrants needed to bring levels of inbreeding down to 0.1 in fragments with previous inbreeding levels of 0.1 and higher (Frankham et al. 2017, Fig. 12.3).

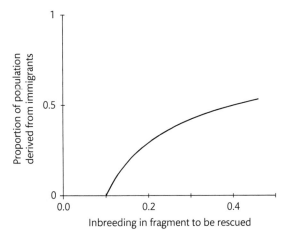

Example 8.1 Calculating the number of immigrants needed to reduce inbreeding

(Frankham et al. 2017, Example 12.2)

If a fragment of size 50 with an inbreeding level estimated to be $F_I = 0.2$ is augmented with unrelated immigrants, the proportion of non-inbred immigrants (X) needed to reduce the level of inbreeding to $F_{Pooled} = 0.1$ is:

$$X = 1 - \sqrt{F_{pooled}/F_I} = 1 - \sqrt{0.1/0.2} = 0.29$$

Thus, 29% of the pooled population needs to derive from immigrants, or ~ 15 immigrants that breed.

As a general rule, the numbers of immigrants will need to be higher if the source population is partly inbred or related to the recipient population. Overall, it is better to begin augmentation when the population is only moderately inbred.

Practical constraints mean that genetic swamping is unlikely to result from genetic rescues

Very high levels of gene flow can lead to dilution of local genetic composition, or even full replacement. However, genetic swamping should not be an issue when donor and recipient populations are adapted to similar environments. Practical constraints will typically lead to genetic management with minimum acceptable rates of gene flow.

If populations remain small and isolated after genetic rescue, they will need additional augmentation in the future

Unless natural gene flow can be restored, genetic rescue may have to be undertaken repeatedly to ensure the survival of genetically isolated populations. For example, the

Illinois population of greater prairie chickens was genetically rescued but is again becoming inbred (Chapter 5 frontispiece; Mussman et al. 2017). Genetic rescues will typically be required less frequently if the drivers of decline have been removed and the size of the rescued population increases.

Lower immigration rates, smaller N_e, and inbred donors all mean that more frequent translocations are needed. For example, completely isolated population fragments of effective sizes 20, 50, and 100 require augmentation of gene flow after approximately 2, 10, and 21 generations, respectively to avoid losing more than 10% of their genetic diversity (Example 8.2). These N_e correspond to populations with ~ 160, ~ 400, and ~ 800 breeding individuals for a species with an average N_e/N ratio of 0.125, and related adjustments can be made for species with other ratios. Given the number of variables, we recommend that computer modeling, akin to that done for the Allegheny woodrat using VORTEX, be used to predict the number of generations until augmentation of gene flow is needed.

> The frequency of translocations between donor and inbred recipient populations depends on the gene flow rate, the effective size of the recipient population, and whether the donor population is outbred or inbred

---

Example 8.2 Generations until an additional round of gene flow is required into completely isolated population fragments

The number of generations (t) until another round of gene flow is required for completely isolated population fragments depends on the effective population size (N_e) and the reduction in heterozygosity. If the first round of gene flow is done when a population has lost 10% of its genetic diversity and it is crossed to another equally inbred population, the resulting population will only have its heterozygosity increased to 0.95 that of the initial population. Under these circumstances, the heterozygosity when a second genetic rescue was required would be 0.9/0.95 ~ 0.947 relative to the heterozygosity in the initial population before its decline ($H_{threshold}$).

We can estimate the approximate number of generations until another round of gene flow is required for a population of size $N_e = 20$ as:

$$t = -2N_e \ln\left(H_{threshold}\right) = -2 \times 20 \ln\left(0.947\right) = 2.2 \text{ generations}$$

For populations of size $N_e = 50$ and 100, the values are 5 and 11 generations, respectively. If the initial augmentation involved crossing an inbred population to a fully outbred population, the number of generations until a second round of augmentations is required with N_e of 20, 50, and 100 are 3, 8, and 16 generations, respectively.

If needed, a second genetic augmentation should if possible use a different isolated donor population (or populations) than the initial one, as this should be more beneficial than using the same one again.

Rotational movement of individuals between multiple fragments is strategy worth considering

A simple strategy for managing gene flow between multiple fragments is using regular exchange of individuals between fragments in a rotational manner (Frankham et al. 2017). In the first generation of management, individuals are moved from their natal fragment to another fragment (from fragment 1 to 2, 2 to 3, etc.). In the subsequent generation, individuals are exchanged between the next set of fragments (from fragment 1 to 3, 2 to 4, etc.). The pattern will depend on the number of fragments, but for any set of fragments a scheme can be developed to ensure the fragments contain an admixture of individuals from other fragments after a small number of generations. This pattern is then repeated over generations. We recommend consideration of this strategy for management of fragmented populations.

We now turn to management of gene flow based on detailed genetic information.

Management of gene flow with detailed genetic information

Data on genetic diversity from molecular genetic and genomic analyses allow more precise genetic management, leading to better choice of donor populations, higher genetic diversity, and lower inbreeding than with management based on less information

Box 8.2 provides a case study of genetic management of population fragments of the golden lion tamarin, including wild, translocated, captive, and reintroduced populations, with part of it based on detailed genetic information and part on general principles.

Box 8.2 Genetic management of wild, captive, and reintroduced populations of golden lion tamarins

(after Frankham et al. 2017, p. 266 and Box 13.1)

Golden lion tamarins are small, arboreal, monogamous primates from Brazil (see chapter frontispiece) that became endangered following habitat reduction and fragmentation in the Atlantic Rainforest to less than 2% of its original area. The Golden Lion Tamarin Conservation Program integrates captive breeding, reintroductions, translocations, habitat restoration, research, community education, and outreach.

The wild population has recovered from < 150 to ~ 3,200 since 1970 as a result of integrated conservation management. Based on general principles, translocations of all animals from the most severely threatened populations were used to establish a new protected population with enhanced genetic diversity (fragments 1 and 2 on the chapter frontispiece map), now numbering about 130 tamarins. Fragments 3, 5, 8, 9, 10, and 11 (lightly shaded) were founded and managed with gene flow from the captive population. A total of 153 tamarins were released between 1984 and 2001, and the reintroduced population has flourished, currently numbering over 750. The captive population was managed by minimizing mean kinship based on pedigrees, and kinship information was also used to select individuals for reintroduction so that most of the captive population's founder genetic diversity was transferred to the wild population.

Census information is used to design the program of regular translocations among fragments to minimize inbreeding and maximize the effective size of the entire population. Further, habitat corridors have been planted to reconnect the largest forest fragments.

Before we can address detailed management of gene flow, we must first determine what genetic parameter to manage.

Minimizing mean kinship is superior to alternatives for managing gene flow between fragments

Minimizing mean kinship (mk) is the recommended procedure for managing pedigreed populations of threatened species in captivity to maximize retention of genetic diversity, based on computer simulations and analytical and empirical studies.

Minimizing mean kinship is also recommended for managing gene flow between population fragments in the wild. It is superior to F_{ST} (sometimes recommended for use in conservation management) because F_{ST} and related parameters are extremely sensitive to levels of genetic diversity in the populations, preventing them from reliably reflecting the relatedness of individuals among populations (Box 8.3). Populations with the highest F_{ST} to a target managed population will usually have low genetic diversity themselves (Chapters 4 and 7): a more effective choice would be to use a more genetically diverse population as the source for genetic rescue.

Box 8.3 Managing by mean kinship is superior to managing using F_{ST} and other metrics

(after Frankham et al. 2017, Box 13.2, utilizing data from Culver et al. 2000)

We compared the ability of mk versus F_{ST} to identify the optimal source population for restoring genetic diversity to the depleted Florida panther population using data on allele frequencies at ten microsatellites in six regional populations of puma (*Puma concolor*) across the Americas.

Region	Number of alleles/marker	Heterozygosity
FL (Florida)	1.2	0.04
NA (North America)	6.4	0.55
CA (Central America)	5.4	0.67
NSA (Northern S America)	8.7	0.80
ESA (Eastern S America)	8.3	0.76
CSA (Central S America)	6.7	0.76
SSA (Southern S America)	5.8	0.67

Florida panther (USA)

The following matrix contains the pairwise values of mean kinship between populations (below diagonal, high values indicate low divergence) and pairwise F_{ST} (above diagonal, high values indicate high divergence). The final row of the table shows the heterozygosity of the FL population prior to genetic rescue and the heterozygosities that would result if 20% of the FL population were replaced by immigrants from each of the other populations.

Continued ➤

	FL	NA	CA	NSA	ESA	CSA	SSA
FL		**0.245**	0.338	0.327	0.270	0.332	**0.350**
NA	0.452		0.087	0.113	0.089	0.133	0.163
CA	0.234	0.275		0.075	0.076	0.092	0.116
NSA	**0.156**	0.161	0.144		0.089	0.133	0.163
ESA	0.278	0.217	0.164	0.159		0.035	0.085
CSA	0.184	**0.149**	0.142	0.158	0.184		0.044
SSA	0.212	0.152	0.156	0.126	0.151	0.219	
Resulting H	0.043	0.225	0.299	**0.330**	0.289	0.319	0.307

Using *mk*, NSA was identified as the best source to restore genetic diversity to the Florida population: it is the most genetically diverse population as well as being most divergent from FL (*mk* = 0.156, shaded, in the matrix table). A population with 20% of its genes from NSA and the rest from FL would increase heterozygosity from 0.043 to 0.330 (bold), higher than any of the alternative augmentations.

By contrast, F_{ST} identified SSA (0.35, shaded) as the most divergent population and thus the best source of immigrants, but SSA has lower heterozygosity than do the other South American populations, and its use would not increase heterozygosity as much as introducing immigrants from NSA.

We also considered another scenario in which the NA population required genetic rescue. *Mk* identified CSA as the best source population (0.149 in bold), a population with high genetic diversity and considerable divergence in allele frequencies from NA.

Conversely, F_{ST} identified FL as the source for improving diversity in NA (0.245 in bold), even though FL is highly inbred and in need of rescue itself, and contains no alleles not also found in NA. Indeed, F_{ST} erroneously indicated that the highly inbred FL population would be the best choice for genetic supplementation of any of the other populations.

Neither F_{ST} nor any of several other tested metrics consistently perform as well as *mk*. Further, relatedness, a concept that is often confused with kinship, is not appropriate for genetic management purposes, as the two measures differ in inbred populations.

Should we manage to conserve rare alleles?

Genetic management with the primary purpose of conserving rare alleles for individual genes or gene clusters is not recommended

Hughes (1991) suggested genetic management to maximize the conservation of the number of major histocompatibility complex (MHC) alleles, as diversity for those alleles is likely to be beneficial to population persistence. However, many other areas of the genome also contain genetic diversity of functional significance. Further, many rare alleles in the genome are harmful and their frequencies reflect mutation-selection balance, so we do not wish to increase their frequencies.

Proposals to conserve rare alleles of individual genes or gene clusters are not recommended because they result in faster genome-wide loss of genetic diversity and faster rates of inbreeding than managing total genetic diversity in the genome by managing to minimize mean kinship.

What is kinship and how do we manage fragmented populations using it?

Kinship measures how related individuals and populations are

The kinship (or coancestry) between individuals i and j (k_{ij}) is the probability that two alleles of a gene taken at random, one from each individual, are identical by descent, and would be the inbreeding coefficient of any offspring they had. The mean kinship of an individual i (mk_i) is the mean of kinships between that individual and all living individuals in the population including itself. The mean kinship of the population ($\overline{mk}$) is the average of all individuals' mean kinships. As kinship is most easily understood by considering pedigrees, we illustrate estimation of kinship and mean kinship for a hypothetical pedigree with two populations in Example 8.3. However, for most wild populations, kinship will be estimated from microsatellites or SNPs.

Example 8.3 Calculating kinships in two hypothetical population fragments
(after Frankham et al. 2017, Example 13.1)

Consider an original group of six unrelated individuals (top row of circles) that are paired and bred and their descendants (bottom row of individuals) placed into two fragments (A and B).

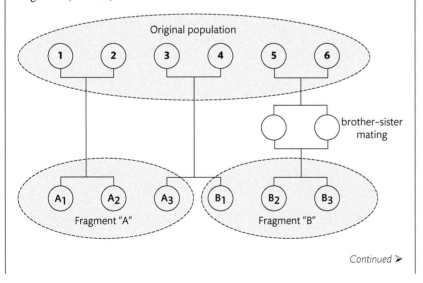

Continued ➤

Kinship matrix

The kinship matrix shows the kinships of all possible pairwise combinations of individuals, based on the pedigree above. Note that the inbred individuals (B_2 and B_3) have self kinships greater than those for non-inbred individuals (0.625 vs 0.50). In the table below, within fragment kinships are shown in light shading, while between population kinships are unshaded.

		Individuals					
		A1	A2	A3	B1	B2	B3
Individuals	A1	0.5	0.25	0	0	0	0
	A2	0.25	0.5	0	0	0	0
	A3	0	0	0.5	0.25	0	0
	B1	0	0	0.25	0.5	0	0
	B2	0	0	0	0	0.625	0.375
	B3	0	0	0	0	0.375	0.625

Calculating within population mean kinships for population A (mk_{AA}) and B (mk_{BB}), we obtain (ignoring zeros):

$$mk_{AA} = \frac{0.5 + 0.5 + 0.5 + 0.25 + 0.25}{9} = 0.22$$

$$mk_{BB} = \frac{0.5 + 0.625 + 0.625 + 0.375 + 0.375}{9} = 0.28$$

while the between population kinship (mk_{AB}), as discussed in the following text, is:

$$mk_{AB} = \frac{0.25}{9} = 0.028$$

The mean kinship between populations A and B (mk_{AB}) is the average of the pairwise kinships of all individuals in population A with all individuals in population B, as illustrated in Example 8.3 (note that $mk_{AB} = mk_{BA}$). We will illustrate use of mean kinship in management of genetic rescues below.

Why is mean kinship superior to other metrics for managing genetic diversity?

The average mean kinship for a population ($\overline{mk}$) is the expected inbreeding coefficient in the next generation with random mating, and is directly related to the proportion of genetic diversity lost since generation 0. Consequently, **if kinship is minimized, inbreeding in the next generation is minimized and retention of heterozygosity is maximized.**

Restoring genetic diversity by gene flow based on minimizing mean kinship between populations

Mean kinships within and between population fragments can be used to identify the best sources of individuals for augmentation. The most genetically valuable single source population of immigrants will possess the lowest between population mk with that fragment. In Example 8.4 below, the Mt. Jasmin and Mt. Sebert fragments have high mean kinships ($\overline{mk}$ = 0.57 and 0.41, respectively) and need augmentation. Bernica has the lowest mean kinship with both of them (mk_{AC} = –0.02 and –0.04) and is the most suitable donor population.

How can we estimate population mean kinship for wild populations?

Mean kinships among populations from historical population sizes and migration

When detailed historical demographic information is available on fragments or individuals, computer modeling can be used to estimate kinships between and within fragments to guide management of gene flow more precisely. For example, if such information was available, genetic management of the Allegheny woodrats in Box 8.1 could be improved, compared to random translocations of individuals among fragments. Ideally, we would move animals from populations with high genetic diversity (low within population mean kinship) into populations with low genetic diversity (high within population mean kinship), and we would move them between populations with the lowest between population mean kinship to achieve maximum genetic benefits.

Mean kinships within and among populations from molecular markers

Mean kinships can be estimated from sharing of alleles by individuals for microsatellites or SNPs. Example 8.4 illustrates the use of microsatellite data to estimate mean kinships within and among populations of the jellyfish tree. Similarly, kinships within and among giant panda populations have been estimated from data on 150,000 SNPs.

Various software packages are available to estimate kinships from molecular data: one is listed in the Software section at the end of the chapter, and new tools are produced frequently, especially within the R computing environment. The probability of detecting kinship differences depends on the number and kind of markers used, number of individuals sampled per population, and the degree to which populations differ in mean kinship. Tools exist for planning sampling strategies to detect population structure based on different types of markers (Hoban et al. 2013; Flesch et al. 2018). For our purposes, we need to be able to detect differences in mean kinship at a level of 0.10 or finer. In small populations with low genetic diversity, more markers are usually required for useful kinship estimation, while fewer are needed for larger populations with higher genetic diversity.

However, given the much higher precision for estimating kinships using thousands of SNPs compared to tens of microsatellites, we recommend that 10,000 or more SNPs now be used instead of microsatellites (Weir & Goudet 2017; Goudet et al. 2018).

Example 8.4 Within and between population kinships in the jellyfish tree
(after Frankham et al. 2017, Example 13.3, based on Finger et al. 2011)

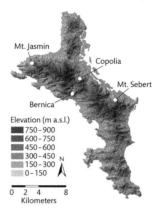

The mean kinships between and within the four populations of the critically endangered jellyfish tree on the island of Mahe in the Seychelles were determined from genotypes of all 90 existing trees for ten microsatellite markers (see following table).

Kinships (k_{ij}) were estimated from correlations of allele frequencies among all individuals, with standardization and adjustment for bias. Mean kinships within and between populations were then computed (see values in kinship matrix table).

Population (N)	Mt. Jasmin	Copolia	Mt. Sebert	Bernica
Mt. Jasmin (2)	0.57	0.07	0.02	−0.02
Copolia (3)	–	0.17	0.09	−0.02
Mt. Sebert (7)	–	–	0.41	−0.04
Bernica (78)	–	–	–	0.004

The largest population, Bernica, had much lower within population mean kinship (0.004) than did the other populations, and low between population kinships with the other populations. Thus, genetic rescue would best be accomplished by gene flow from Bernica to the other populations. A cross between Bernica and Mt. Sebert populations resulted in a 151% improvement in fitness, compared to the inbred Mt. Sebert population.

Jellyfish tree (Seychelles)

More precise management is possible when kinships are available for all individuals

Information on kinships between individuals within and among fragments allows managers to identify individuals best suited to transfer—those least related on average to the recipient population—in addition to identifying the best source population.

Optimal genetic rescue is identified by evaluating what each individual's mean kinship would be if it were transferred to each fragment, and then sending each individual to the recipient fragment to which it has the lowest mean kinship. This is being done when deciding which California condor chicks hatched in captivity should be transferred to each of three wild populations.

Individual kinships can be estimated with high accuracy from genomic information. However, we recommend that individual level management in the absence of pedigrees be avoided unless many thousands of SNPs are used to estimate kinships on all individuals in natural populations (Nietlisbach et al. 2017).

California condor (USA)

When to cease?

Reasonable objectives would be to cease augmentations when heterozygosity is > 95% of the maximum achievable for the target population (given the available donor sources) and the inbreeding coefficient is less than 0.05. Additional augmentation may be required later if a fragment continues to have a small population size with consequent loss of genetic diversity (see above).

Populations should be monitored to confirm that augmentation has resulted in intended benefits

It is important, if possible, to monitor the long-term consequences of gene flow between fragmented populations to quantify:

- the demographic response of populations to augmentation
- levels of realized gene flow compared to targets
- whether there is some natural gene flow from nearby populations, undocumented translocations, or other species
- whether loss of genetic diversity is faster than expected
- whether natural selection is influencing the impacts of augmentation
- how rapidly inbreeding increases subsequent to a genetic rescue.

Genetic determinations pre- and post-augmentation using microsatellites or SNPs have been done, for example in Mauritius kestrels, Florida panthers, greater prairie chickens, Isle Royale gray wolves, Scandinavian wolves, European adders, Trinidadian guppies, and Columbia Basin pygmy rabbits (DeMay et al. 2017). These examples encompass various combinations of molecular marker data, adaptive variation, fitness, and demography. However, few such studies have been done in wild invertebrates or plants.

Mauritius kestrel (Mauritius)

Integrating genetic rescue with other management considerations

There are often behavioral, ecological, demographic, disease, and political considerations, and other genetic issues not associated with fragmentation to consider in programs that manage gene flow. For example, when new individuals are introduced into a population, introduced animals can be killed or maimed by residents, or new males may kill resident juveniles, as in lions. We recommend integrated consideration of all concerns, as encompassed in the One Plan Approach, or others based on similar principles, but details are beyond our remit (Byers et al. 2013).

In the next chapter we address the increased need for genetic management due to the rapid environmental changes resulting from global climate change.

Summary

1. Sound management strategies for gene flow into populations suffering genetic erosion can be instituted without detailed genetic data, although additional demographic information and modeling can provide information for more precise and effective management.
2. When limited information is available, the default should be to initiate genetic augmentation based on general principles rather than doing nothing. Donor and recipient populations should be largely isolated from each other, large donor populations are better than smaller ones, outbred populations better than inbred ones, and several isolated donor populations are better than one. Choice of individuals to transfer from donor into recipient populations should minimize the chance of selecting close relatives, and they should be taken from locations throughout the donor's population range to capture maximum genetic diversity.
3. Any appropriate gene flow into an isolated inbred population is better than none, but we recommend at least five genetically contributing immigrants per generation to avoid accumulation of damaging levels of inbreeding and to increase adaptive potential.
4. With detailed genetic information, minimizing mean kinship is the best approach to genetic management; using F_{ST} is inappropriate.
5. Mean kinships within and among wild populations can be estimated at the population and individual level from modeling or genotypic data for many genetic markers. Individuals with low mean kinship to the recipient population have the best potential to increase genetic diversity and reduce inbreeding.
6. Populations should be monitored to ensure that movement of individuals or gametes has resulted in the desired demographic and genetic outcomes.

FURTHER READING

Ballou & Lacy (1995) Pivotal paper justifying the use of minimizing mean kinship to manage threatened populations in captivity based on analytical work and computer simulations.

Byers et al. (2013) Describes the One Plan Approach that encompasses all issues in the management of threatened species.

Frankham et al. (2017) *Genetic Management of Fragmented Animal and Plant Populations*: Chapters 12 and 13 have more detailed treatment of topics in this chapter.

Goudet et al. (2018) Discusses how to estimate kinship from SNP data and evaluates different methods, and the number of SNPs required for reliable estimates.

IUCN/SSC (2013) *Guidelines for Reintroductions and Other Conservation Translocations*: provides useful guidelines for translocations.

SOFTWARE

HIERFSTAT: a package that can be used for estimating kinships from genetic marker data using the r^B method of Weir & Goudet (2017), the method we currently recommend (Goudet 2005 and its update). https://github.com/jgx65/hierfstat

METAPOP: software for the dynamic and flexible management of fragmented populations (Pérez-Figueroa et al. 2009). https://gitlab.com/elcortegano/metapop2

NEWGARDEN: a program to model the growth and population genetics of plant populations developing from different founding and life history conditions (Pelikan & Rogstad 2013). http://homepages.uc.edu/~pelikas/NEWGARDEN/index.html

VORTEX: population viability analysis software that tracks the predicted heterozygosity and inbreeding coefficients of individuals, and thus can be used in management of gene flow when there is limited genetic information (Lacy & Pollak 2014). https://scti.tools/vortex

Global climate change increases the need for genetic management

Adverse genetic impacts on fragmented populations will accelerate under global climate change. Many populations and species may be unable to adapt *in situ*, or to move unassisted to suitable habitat. Genetic management may reduce these threats by augmenting genetic diversity to improve the ability to adapt evolutionarily and reduce inbreeding or by enhancing the success of translocations. Global climate change amplifies the need for genetic management of fragmented populations.

TERMS

Assisted colonization, backcross, *in situ*, invasive species, phenotypic plasticity

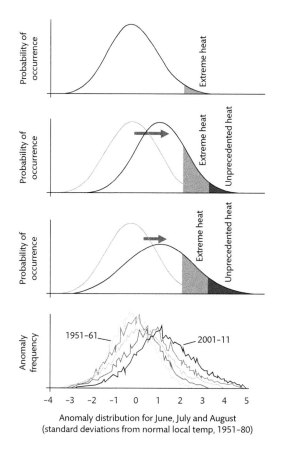

Anomaly distribution for June, July and August
(standard deviations from normal local temp, 1951–80)

Increases in global temperatures and heat anomalies since 1951. The top three panels illustrate the predicted impacts on global climate. The first panel represents pre-industrial levels. The second predicts impacts of an increase in mean temperature, while the third illustrates increases in both mean and variability resulting in even higher frequencies of extreme and unprecedented heat events. The observed temperature distributions from 1951–1961 to 2001–2011 (bottom panel) show increases in means and variability over time (NOAA 2016, http://www.ncdc. noaa.gov/cag/).

A Practical Guide for Genetic Management of Fragmented Animal and Plant Populations. R. Frankham, J. D. Ballou, K. Ralls, M. D. B. Eldridge, M. R. Dudash, C. B. Fenster, R. C. Lacy & P. Sunnucks. Oxford University Press (2019). © R. Frankham, J. D. Ballou, K. Ralls, M. D. B. Eldridge, M. R. Dudash, C. B. Fenster, R. C. Lacy & P. Sunnucks 2019. DOI: 10.1093/oso/9780198783411.001.0001

Global climate change is harming many species and its impacts are projected to worsen

Human activities have resulted in the warming of the planet, increasing frequencies of extreme weather events, rising sea levels, and ocean acidification. For example, extreme high temperature events have increased from covering 0.1–0.2% of the planet in 1951–1980 to 10% in 2006–2010. Further, these effects are projected to worsen, so species face rapidly changing environments (IPCC 2014; Frölicher et al. 2018).

Innumerable terrestrial and marine species across the planet are adversely impacted by global climate change (Scheffers et al. 2016). For example, coral reef communities are severely impacted (Box 9.1). Further, extinctions of two populations of checkerspot butterfly (*Euphydras editha bayensis*) in California, and the Bramble Cay melomys rodent (*Melomys rubicola*) species from Australia have been attributed to climate change, the former due to increased variability in precipitation and the latter to sea level rise.

Brain coral (*Gonisastrea flavulus*)
(Indo-Pacific)

Box 9.1 Adverse impacts of global climate change on coral reef communities

(after Frankham et al. 2017, Box 14.1, based on Hoegh-Guldberg et al. 2007; De'ath et al. 2012; van Oppen et al. 2015)

Corals are suffering increasingly frequent episodes of mass bleaching (first recorded in 1980), killing them and reducing suitable habitat for other reef organisms. For example, the 2016 bleaching on the Great Barrier Reef, Australia was the worst on record and similar impacts were recorded on many other reef systems across the planet (Hughes et al. 2017, 2018). Ocean acidification is also impacting corals and other reef invertebrates and is projected to worsen. Further, storms are damaging reefs weakened by bleaching and ocean acidification. Coral mass has halved on the Great Barrier Reef between 1985 and 2012, and 80% of its corals may be gone within 40 years without world action to reduce carbon pollution.

The impacts of climate change on corals are so extreme that they are expected to outpace the evolutionary capacity of corals to adapt to acidification and warming waters. Consequently, a series of innovative projects is underway in an attempt to save corals:

- translocating better performing corals from regions where they are doing well to regions where they are doing poorly
- introducing zooxanthellae symbionts that confer enhanced coral survival and growth under projected environments
- hybridizing closely related species to determine whether progeny will be more resilient than the parents
- selective breeding to improve their ability to persist in the face of global warming and ocean acidification
- conditioning corals in the laboratory to make them more resilient to stress, and releasing them into the wild.

Global change is accelerating extinction risk for fragmented populations

With global climate change, many species are faced with altered environments to which they are not adapted, with consequent reductions in reproduction and survival. Demographic and environmental variability and catastrophes are worsening, thus accelerating adverse genetic impacts. Many species are being adversely impacted by multiple deterministic and variable factors. For example, the Arctic fox (*Alopex lagopus*) is facing multiple threats to its persistence associated with climate change (Box 9.2), as are many other species.

Box 9.2 The Arctic fox is experiencing multiple threats associated with global climate change

(Frankham et al. 2017, Box 14.2, based on Sillero-Zubir & Aungerbjorn 2009)

The Arctic fox inhabits the Arctic tundra and sea ice. Populations in Sweden, Finland, and Norway were decimated by fur hunting in the early twentieth century and have failed to recover despite legal protection since 1940. They now number only ~ 150 individuals and are threatened with extinction. Although the Arctic fox in the rest of its range is relatively abundant, it has been disappearing from the southern edge of the tundra, raising concerns over the species' long-term future.

The species faces multiple threats from climate change: its tundra habitats are shrinking, the sea ice is disappearing, its lemming prey are becoming less abundant in some areas, and it is being displaced by the larger red fox (*Vulpes vulpes*), which is moving northward as temperatures warm.

A small inbred Swedish population was genetically rescued by natural immigration of three males from Norway, which reduced inbreeding, enhanced genetic diversity, improved fitness, and led to a doubling of its population, which should enhance its ability to cope with threats from climate change (Hasselgren et al. 2018).

Arctic fox

Extinction vortices are accelerating due to worsening deterministic factors, increasing variability, and their interactions because of global climate change.

Global climate change is expected to accelerate the extinction vortex by:

- reducing species' fitness and population sizes as they become less well adapted to their environment
- increasing environmental variability
- increasing frequency of catastrophes
- increasing demographic variability
- increasing genetic problems
- strengthening interactions among these effects.

For example, the last known population of timber rattlesnake (*Crotalus horridus*) in New Hampshire, USA is declining due to the interactions among climate change, disease, and low genetic diversity.

The above factors combine to reduce population size and increase its variability, thus worsening genetic problems from four effects:

- decline in fitness in species that cannot evolve fast enough to keep up with climate change, reducing N_e for many species
- N_e will be a smaller proportion of N due to increasing climatic extremes, because fewer individuals will breed and there will be greater fluctuations in population sizes
- worsening inbreeding depression due to more rapid inbreeding and more stressful environments
- reduced ability of many populations to adapt genetically (due to diminished N_e, loss of genetic diversity, reduced fitness, and increased environmental variability).

These impacts are even more serious in fragmented habitats and fragmentation is expected to worsen as habitat suitability declines. For example, model experiments using rotifers showed that habitat fragmentation and environmental warming stressors caused populations to decline much more rapidly in combination than when acting alone.

How have species responded to these problems?

Species have responded to global climate change by moving, adapting, or going extinct

Moved

Many species have already moved in response to climate change, with range shifts averaging 16.9 km per decade towards the poles. Examples range from mammals to mollusks, and grasses to trees. In 80% of cases, movement has been in the predicted elevational or compass direction.

Adapted via phenotypic plasticity

Species often respond to changed environments via phenotypic plasticity. In plants, this may involve shifts in timing of annual events, or changes in leaf size, shape, thickness, or pigmentation, height at maturity, water use efficiency, seed size and number, chemical defenses, etc. Similarly, animals show many plastic responses to environmental changes, including in behavior, timing of migration, breeding time, and body weight.

Many species have shown changed timing of annual events, such as earlier flowering and fruiting in plants, emergence of butterflies, breeding in frogs, and migration in some birds. For example, 43 plant species are flowering on average 7 days earlier than in the 1850s at Concord, Massachusetts, USA in response to warming. However, pheno-typic plasticity may not produce the large and persistent inherited changes quickly needed to cope with ongoing directional climate change.

Adapted genetically

Many species have evolved in response to altered climatic conditions. These include Canadian red squirrels (*Tamiasciurus hudsonicus*), blackcap warblers (*Sylvia atricap-illa*), several species of *Drosophila* fruit flies, harlequin fly (*Chironomus riparius*), pitcher plant mosquitoes (*Wyomia smithii*), Mediterranean wild thyme (*Thymus vulgaris*), canola (*Brassica rapa*), and thale cress (Waldvogel et al. 2018). For example, average flowering times of wild populations of barley (*Hordeum spontaneum*) and emmer wheat (*Triticum dicoccoides*) in Israel evolved to be 8.5 and 10.9 days earlier, respectively, in plants derived from seed collected in 2008 than in those from 1980 seed (Fig. 9.1).

Emmer wheat (Israel)

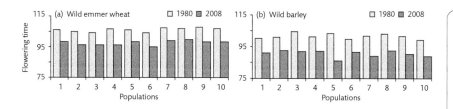

Fig. 9.1 Evolution of earlier flowering time in wild populations of (a) emmer wheat and (b) wild barley in Israel in response to global climate change. Differences in days from germination to flowering for ten populations of each species derived from seed collected from the wild in 1980 and 2008, and grown contemporaneously in a greenhouse under the same conditions (Frankham et al. 2017, Fig. 14.2, based on Nevo et al. 2012).

What are the consequences of global climate change for conservation management?

Many species will require assistance to persist under climate change because they will be unable to adapt genetically or move to more suitable locations

In many cases, changes in environments are projected to occur at rates greater than movements of species reflected in paleoecological records. Some populations are evolving too slowly to keep pace with environmental changes, for example great tits and European beech, likely causing loss of species from local biota. For species with small geographic ranges or narrow habitat tolerances, the result can be complete extinction of the species, especially for endemic species on small islands (e.g. Lord Howe Island woodhen [*Hypotaenidia sylvestris*], Australia; Santa Fe land iguana [*Conolophus pallidus*] and several species on the Galápagos islands, Ecuador; and the Catalina mahogany tree [*Cercocarpus traskiae*], California), species restricted to mountain tops (e.g. mountain pygmy possums), and critically endangered species with a single population, or a few nearby ones (e.g. southern corroboree frog and Wollemi pine, Australia).

In this chapter, we consider how the challenges from climate change alter our genetic management recommendations for fragmented populations. We conclude that the genetic management principles discussed in previous chapters still apply, but the need for active genetic management of fragmented populations is becoming ever more urgent because of global climate change.

We next consider how genetic management might help save populations and species that cannot currently adapt to climate change in their current location or move to other suitable locations.

Genetic management options for populations that cannot move or adapt sufficiently

A decision tree for assessing genetic management options to improve the ability of a population to cope with climate change is given in Fig. 9.2.

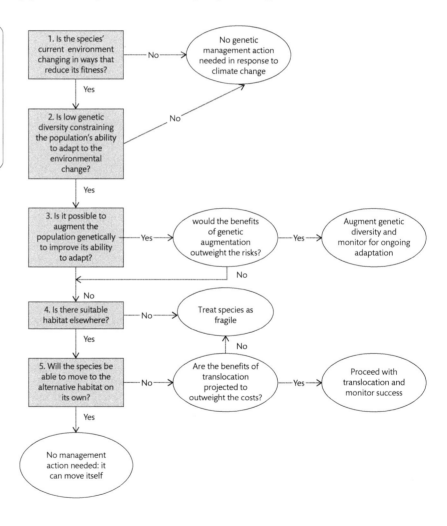

Fig. 9.2 Decision tree for assessing genetic management options to improve the ability of populations and species to cope with global climate change (after Frankham et al. 2017, Fig. 14.3). Although the tree shows genetic augmentation and translocation being decisions on alternate branches, neither strategy is guaranteed of success. It will often be wise to pursue both.

Given the effects of climate change on species, active management will often be needed to prevent population or species extinctions by:

- improving the ability of species to adapt evolutionarily
- translocation to more suitable habitats
- both.

The first step is to identify populations that are incapable of adapting rapidly enough or moving.

Identifying populations incapable of adapting rapidly enough or moving

The most comprehensive approach to identifying species at risk from climate change combines evidence from:

- direct observations
- paleoecological records
- ecophysiological models
- climate envelope models
- population models
- experimental manipulations.

We endorse the approach of using all available evidence, but emphasize the need for prompt action to address threats to species persistence. Gaps in knowledge can then be progressively filled and incorporated in an adaptive management process.

The IUCN Climate Change Specialist Group recently published guidelines for assessing the vulnerability of species to climate change (see also Foden et al. 2018 and Razgour et al. 2018). The most comprehensive single means for assessing whether a species needs translocation is to conduct computer simulations using spatially explicit models that incorporate projected global climate change and genetic factors. For example, modeling was used to identify populations of bull trout (*Salvelinus confluentus*) that are vulnerable to climate change. However, detailed individual species modeling typically requires extensive species-specific biological, ecological, and genetic data that are often not available.

Consequently, simpler climate envelope models have been used to identify species at risk from climate change. Using this method, an estimated 35% of species on the planet are on a path to extinction under the global climate change scenario that the planet is currently tracking or exceeding.

A simpler assessment indicates that the Florida torreya tree would likely benefit from translocation, as fewer than 1,000 individuals of this long-lived species exist in a restricted area, it is not reproducing in its native habitat, is unlikely to move naturally, has low genetic diversity, and is unlikely to adaptively evolve *in situ*.

Florida torreya (USA)

We next deal with means to improve the ability to adapt evolutionarily. Populations that are unlikely to be able to adapt evolutionarily will have some combination of the following attributes:

- experiencing rapid climate change
- small population size
- low genetic diversity
- low N_e
- low reproductive rates
- long generation intervals.

For example, the European beech and other trees often have limited ability to rapidly adapt because of long generation times. Even some species with rapid generation times cannot adapt due to the lack of adaptive genetic diversity, including tropical *Drosophila* species that cannot evolve desiccation resistance. Similarly, yellow warblers (*Setophaga petechia*) are failing to adapt rapidly enough to prevent population declines, due to insufficient adaptive genetic variation (Bay et al. 2018). Further, wild mustard (*Brassica juncea*) was unable to adapt to projected global climate change conditions, as inbreeding depression worsened as the projected environments became increasingly stressful. Conversely, calcifying organisms, such as sea urchins, bryozoans, and coccolithophores (unicellular algae), presumed to have high population sizes, ample genetic diversity, and short generation intervals, can adapt evolutionarily in response to increased ocean acidification.

PVA simulations, run with and without evolutionary adaptation, provide an important means for identifying populations incapable of keeping pace with climate change. For example, with a simulated increase in summer temperatures of 2° C by 2100 for sockeye salmon (*Oncorhynchus nerka*) in the Fraser River, Canada, adult migration from the ocean to the river was predicted to advance by ~10 days due to evolutionary change, substantially reducing the risk of population extinction compared to that without evolutionary change. Few such computer modeling studies encompassing evolutionary adaptation have been conducted, but we recommend them, if feasible.

We next deal with means to improve the ability to adapt evolutionarily.

Implementing genetic management to assist evolutionary adaptation

Even where information on adaptive genetic diversity is lacking, increasing overall genetic diversity can often increase the ability to evolve. For example, studies with *Tribolium* flour beetles and *Drosophila* fruit flies show that gene flow from other populations enhances evolutionary adaptation under changing environments (Chapter 5).

The principles we discussed in Chapters 7 and 8 indicate how to implement such genetic rescues.

From where can adaptive genetic diversity be sourced?

Potential sources of genetic diversity to improve ability of a population to evolve are:

- other populations within the species
- other species (introgression)
- gene editing and transfer.

Up to this point we have focused on within species management. However, when the prospects of saving species this way are very poor, more innovative measures need to be considered. We now consider sourcing new genetic diversity from other species, or creating it using gene editing, given the rapid shifts expected with global climate change.

Introgression from other species is sometimes the only potential source of adaptive genetic variation

Introgression from other species may seem a radical suggestion, as presently even genetic augmentation from other sub-species has typically been avoided by authorities, such as the US Fish and Wildlife Service. However, if there are no other conspecifics to cross with the recipient population, adding new adaptive material by introgression from a related species demands serious consideration for several reasons. First, gene exchange between some species is equivalent to gene flow between populations in others (Chapter 6). Second, many species show low natural levels of introgression (Pennisi 2016). Third, natural utilization of adaptive variation derived from a related species has been reported in butterflies, birds (including Darwin's medium ground finches), primates, and plants (Grant & Grant 2017).

Fourth, some species have evolved by hybridization between species, but remained diploid, as for example in several sunflower species and common ravens (*Corvus corax*) (Kearns et al. 2018). Fifth, multiple species composites have often been used by animal and plant breeders to develop populations for human use in food and horticulture. For example, several cattle breeds were synthesized with contributions from *Bos indicus* and *B. taurus* so they are productive and can tolerate harsh tropical environments.

Gene editing and transfer

Given the severity of the problems that global climate change poses, gene transfer and especially gene editing offer the potential for introducing new adaptive genetic variants. Beneficial alleles from any species can be transferred into a species of interest (without introducing whole genomes), as has been done for example by plant and animal breeders. Whilst there may be reservations about using gene transfer in conservation contexts, there are regulations and protocols governing the creation and release of genetically modified organisms in many countries. The potential benefits of gene transfer are illustrated by the case of the American chestnut (*Castanea dentata*) where a transferred wheat gene confers resistance to an introduced fungus that almost eliminated this iconic tree (Popkin 2018).

Many candidate genes affecting a variety of traits have already been identified as a result of genome sequencing and analysis. For example, a heat tolerant allele of the rubisco activase gene in a wild rice from northern Australia (*Oryza australiensis*) has

Darwin's medium ground finches (Ecuador)

Wild rice (Australia)

been identified that confers tolerance to high temperatures well above the normal optimum and confers heat tolerance when transferred to other species, including domestic rice (Scafaro et al. 2016; B.J. Atwell, pers. comm.). While this gene has not been used for conservation purposes, it represents a promising avenue.

Importantly, CRISPR-Cas9 and related gene editing technologies allow changes to be made to existing alleles without transferring genes from a different species, and has the potential to generate new adaptive genetic variants (Novak et al. 2018).

Other forms of human-assisted evolution

Artificial selection, manipulation of symbionts and other associated microbes, and hybridizations have the potential to improve adaptation in species affected by global climate change.

All these approaches are being tried in corals (Box 9.1).

Below we address genetic management of translocations to cope with climate change.

Genetically managing translocations to cope with climate change

The objective of translocation is to successfully establish or augment populations so that they have a high probability of persisting over the long term. The questions we need to answer are:

- Where should the translocatees be moved to?
- What genotypes or populations should be moved?

Populations not well adapted to their current location are candidates for movement to locations where they are likely to be better adapted now and in the future. This may involve movement to locations outside their historical range.

Founding translocated populations outside the historical range

Moving species beyond their historical ranges is an option for species at immediate risk of extinction and has been undertaken as a last resort in some instances. For example, "species climate" models were used to choose suitable sites for introductions ~ 65 and ~ 35 km, respectively, beyond the then range margins of marbled white (*Melanargia galathea*) and small skipper (*Thymelicus sylvestris*) butterflies in northern England. Translocations were done to these sites in 1999–2000 and both introduced populations grew and expanded their ranges over 6 years (2001–2006), and were still thriving in 2008.

Given that assisted colonization may be the only option to avert extinction under environmental change, we endorse consideration of its use. Its potential use should be carefully evaluated in cost–benefit analyses for each particular case, including the potential for negative effects on existing species in the proposed translocation site.

Choosing populations for translocations

The likely success of a translocation will be greatest when the source population(s) have high genetic diversity, high reproductive fitness at the translocation site, and are already adapted to the environment, characteristics that are rarely found in likely candidates for translocation, unless population crosses are used.

Care must be taken with translocations to ensure that the translocated population is representative of the genetic diversity in the source population(s). An example of poor choice of source population is provided by the koala (*Phascolarctos cinereus*) in southeastern Australia, where ~ 10,000 individuals were sourced from an isolated island population founded with two or three individuals, and moved to other islands and the mainland. The recipient populations now have an inbreeding coefficient of ~ 40% (Johnson et al. 2018). In general, care should be taken when island populations are being considered as source populations for translocation, unless they are crossed with other populations, because they typically have low genetic diversity and are inbred compared to mainland populations.

The ideal population(s) to use for assisted colonizations depend upon the genetic diversity in the population fragments, extent of fragmentation, adaptive differentiation among fragments, and their adaptation to the new site. Figure 9.3 illustrates a range of situations reflecting these issues in the face of global climate change. Where there is evidence of adaptive genetic differentiation among extant populations, the translocated individuals should come from populations most likely to be adapted to the reintroduction habitat, other things being equal. Scenarios 3 and 4 are most relevant to our concerns about fragmented populations. **If the source populations have low genetic diversity and are fragmented, then genetic material for translocation should be sourced from multiple populations and matched ecologically with the intended translocation site** (Fig. 9.3, scenario 4).

Koalas (Australia)

	Genetic diversity of source population	Distribution of source population	Solution	Schematic
1	High	Continuous	Take from entire distribution, or tip if this matches target area ecologically	
2	High	Fragmented along cline	Take from fragment(s) that matches target area ecologically; this might be closest population (particularly in altitudinal series)	
3	High or low	2+ populations- large disjunction- no population adapted to site	Take all populations (especially if low genetic diversity)	
4	Low	Fragmented along cline	Take from multiple populations to augment diversity and adaptation; match ecologically if possible	

Fig. 9.3 Genetic considerations in establishing populations outside their current or historical species range. Relevant scenarios depend particularly on whether levels of genetic variability in the potential source populations are high (H) or low (L), whether populations are fragmented, and whether populations are distributed along environmental gradients (Frankham et al. 2017, Fig. 14.4, after Weeks et al. 2011). The black arrows indicate the recommended sources of individuals to found the population in the new location (gray).

If the new site has an environment that is significantly different from the current site in several attributes (e.g. it is heavily altered by human activities), then establishing the new population with maximum genetic diversity and minimum inbreeding by crossing between several available populations should maximize the probability of success (Fig. 9.3 scenario 3). In many cases, the new habitats will only partially match the environmental requirements of the species, so augmenting genetic diversity to enhance the ability of the population to adapt to its new environment will often be the desired option.

Risk assessment should be undertaken for genetic translocations

A simple risk assessment framework for genetic translocations is given in Fig. 9.4. It is based on whether the translocation is planned within the historical range of the species or outside, genetic structure and isolation of populations, and the likelihood of hybridization to other species or sub-species. An application of this decision tree is given in Box 9.3.

Fig. 9.4 Simplified decision tree for determining whether to proceed or assess risk in translocations (Frankham et al. 2017, Fig. 14.5, after Weeks et al. 2011).

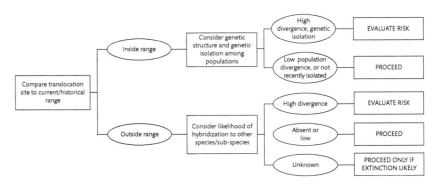

Mountain pygmy possum (Australia)

Box 9.3 Risk assessment for restoring genetic diversity in the mountain pygmy possum (*Burramys parvus*)
(Frankham et al. 2017, Box 14.4, after Weeks et al. 2011, 2015, 2017)

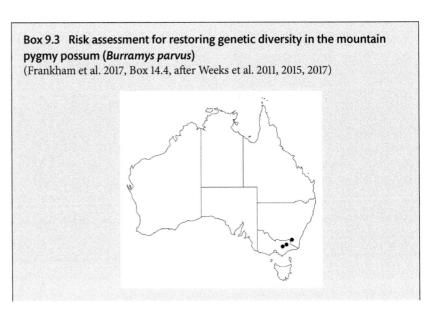

The endangered mountain pygmy possum is restricted to three genetically distinct populations in the Snowy Mountains of southeastern Australia (map), and one of the threats to this species is climate change. The southern population (Mt Buller) crashed in the 10 years subsequent to 1996, losing ~ 70% of its microsatellite genetic diversity and becoming inbred. Consequently, a risk assessment was done for translocating individuals from the central region into the Mt. Buller population to genetically rescue it.

Using the decision tree (Fig. 9.4), the proposed augmentation of the Mt Buller population is inside the species' current range, genetic structure is high among populations, and the four risks needing to be evaluated were outbreeding depression, loss of local adaptation, replacement of the Mt Buller genome with the central genome, and disease transmission. The risk of outbreeding depression was considered moderate, given that these populations have been evolving independently in similar environments for ~ 20,000 years. A mitigation strategy would be to evaluate preliminary crosses, either in the field or in captivity. The risk of losing local adaptation is moderate (in the absence of direct knowledge) and a mitigation strategy would be to backcross to the recipient population. The likelihood of replacing the Mt Buller genome with the central genome was low and could be prevented by translocating only males, and the frequency of gene flow into the population could be monitored genetically. While introduction of a disease could cause extinction of the Mt Buller population, the likelihood of this was minimal given that there are no known diseases present in the possum populations. A mitigation strategy would be veterinary examination of the translocated animals and quarantine if appropriate.

Had the outcome of the decision tree been to abandon the translocation, the risks for the Mt Buller population would have been continued loss of genetic diversity, increasing inbreeding depression, and a high extinction risk, given the population size then was only ~ 30 individuals. The only mitigation strategy would have been to undertake captive breeding to improve the chances of short-term persistence, but genetic adaptation to captivity would become a problem over generations.

In 2010 the Mt Buller population was augmented with males from the central population. Fertile F_1 hybrid individuals have been produced, they have produced offspring, the genetic diversity has more than doubled, and the population size has increased substantially. There has been no evidence of outbreeding depression to date.

Take home messages for readers are located after the Preface. **Final messages for managers of wild animal and plant populations** are presented after this chapter.

Summary

1. Climate change is accelerating genetic erosion and elevating extinction risks for many species, increasing the urgency of genetic management.
2. Many populations and species will be unable to adapt to new environmental conditions, move to more suitable locations, or both.
3. If populations or species have low adaptive genetic diversity, it will often be possible to improve their ability to evolve by augmenting gene flow from other populations,

sub-species, or species. Introducing an allele by gene transfer or modifying one by gene editing is more precise than introducing whole haploid genomes by introgressive hybridization, and may be preferable if beneficial alleles can be identified.

4. Species needing translocations are those having (a) an inability to move without assistance due to poor dispersal, inhospitable matrix, or long distances to new suitable habitat, (b) inability to adapt genetically, or (c) combinations of these.

5. Individuals translocated to cope with climate change should result in a population with ample genetic diversity and low inbreeding that is well adapted to the new environment and has the ability to adapt evolutionarily.

6. If the species consists of small isolated population fragments with high inbreeding and low genetic diversity, and especially when the new site is different in several attributes to the populations' current environment, the individuals for translocation should be taken from several populations and ideally crossed such that F_1 individuals are translocated.

FURTHER READING

Frankham et al. (2017) *Genetic Management of Fragmented animal and Plant Populations*: Chapter 14 has a more detailed treatment of the topic of this chapter.

Hoffmann & Sgrò (2011) Review of climate change and evolutionary adaptation.

Hoffmann et al. (2015) Provides a framework for incorporating evolutionary genomics into conservation management under climate change.

IPCC (2014) *Report of the Intergovernmental Panel on Climate Change*: Addresses likely scenarios for climate change and their consequences.

van Oppen et al. (2015) Review of innovative genetic approaches being used in an attempt to save corals under global climate change.

Weeks et al. (2011) Review of methods of genetic translocation of animals and plants to cope with climate change.

SOFTWARE

HEXSIM: spatially explicit, individual-based, multipopulation, eco-evolutionary modeling environment in which users develop simulations of wildlife or plant population dynamics and interactions (Schumaker & Brookes 2018). www.hexsim.net

MAXENT: software to define environmental envelopes for species, sub-species, and populations, as part of predicting the impact of global climate change on persistence (Phillips & Dudick 2008). http://www.cs.princeton.edu/~schapire/maxent/

VORTEX: population viability analysis software that projects the demographic and genetics prospects for populations into the future (Lacy & Pollak 2014). https://scti.tools/vortex

Final messages for managers of wild animal and plant populations

- The persistence of species with fragmented distributions is heavily dependent upon active genetic management.
- Passive management of small isolated inbred populations, without considering whether they can be genetically rescued, will very likely result in extinctions.
- If there is a donor population of the same species, adapted to similar environments, and without fixed chromosomal differences, then gene flow into the recipient populations has a high probability of improving population fitness and ability to evolve. Almost any regime of gene flow should be beneficial in such circumstances, thereby reducing unnecessary population extinctions.
- Augmentation of gene flow is likely to be a highly cost-effective management option, improving prospects for species to persist in the face of other threats.
- Rapid environmental shifts due to global climate change are increasing the need for genetic rescue, and the urgency of its implementation in the many cases where it would be helpful.

For an overview of the book, readers are referred to the **Take home messages** located after the Preface.

Should you require assistance to implement genetic management, there are conservation and evolutionary geneticists in universities, botanic gardens, museums, research institutions, zoos, and the Conservation Genetics (http://www.cgsg.uni-freiburg.de/) and the Conservation Planning (http://www.cpsg.org) Specialist Groups of IUCN SSC who can assist, or suggest others who can help. We suggest conducting an internet search for conservation, evolutionary, or population geneticists, and check their CVs to see whether they do quality work in areas appropriate to your practical genetic management issues. The Conservation Connectivity (https://conservationcorridor.org) Specialist Group of IUCN may also be a useful source of information.

We trust that this book will contribute to more proactive and effective genetic management of fragmented populations in particular, and more informed consideration of the importance of conserving genetic diversity in general.

A Practical Guide for Genetic Management of Fragmented Animal and Plant Populations. R. Frankham, J. D. Ballou, K. Ralls, M. D. B. Eldridge, M. R. Dudash, C. B. Fenster, R. C. Lacy & P. Sunnucks. Oxford University Press (2019). © R. Frankham, J. D. Ballou, K. Ralls, M. D. B. Eldridge, M. R. Dudash, C. B. Fenster, R. C. Lacy & P. Sunnucks 2019. DOI: 10.1093/oso/9780198783411.001.0001

Glossary

Revised and updated from Frankham et al. (2004, 2010, 2014, 2017).

Adaptive evolution: Genetic change due to natural selection that improves the fitness of a population in its environment.

Allele: An alternative form of a genetic marker or gene, e.g. copies of a microsatellite with different numbers of repeats of the AC sequence.

Allelic diversity: Average number of alleles per genetic marker or gene, a measure of genetic diversity within a population.

Allopatric: The spatial arrangement of populations or species whose geographic distributions do not overlap.

Allozyme: Alternative form of a protein detected by electrophoresis and protein staining that is due to alternative alleles at a single gene.

Assisted colonization: Human-assisted movement of wild plant or animal individuals to a location outside their historical distribution.

Backcross: Cross of hybrid progeny to one of the parental genotypes.

Balancing selection: Selection that maintains genetic variation in a population, encompassing heterozygote advantage (overdominance), rare-advantage selection, and particular forms of selection that vary over space or time.

Base population: Population from which studied population(s) were founded.

Bayesian: Methods of statistical analysis that incorporate other (prior) information, originally devised by Thomas Bayes.

Biological Species Concept (BSC): Concept that defines reproductively isolated units as species. Gene flow is possible within species, but absent or weak between them.

Bottleneck: A sudden restriction in population size, usually resulting in loss of genetic diversity.

Breeding system: Traits of plants that influence mating patterns. For example, characteristics of a flower that influence gamete transfer among conspecifics, e.g. the spatial and temporal arrangement of male and female reproductive organs within a flower.

Catastrophe: An extreme environmental fluctuation that has a devastating impact on a population, e.g. hurricane, drought, extreme winter, fire, disease epidemic.

Census population size: Number of individuals in a population, often potentially breeding adults (compare effective population size).

Centric fusion: Chromosome rearrangement where two chromosomes (with centromeres at or near one end) have fused.

Chloroplast DNA (cpDNA): Circular DNA genomes found in the chloroplasts of plants. Usually maternally inherited in angiosperms (sometimes biparentally), but paternally in gymnosperms.

Chromosomal translocation: A chromosomal variant where segments of non-homologous chromosomes have swapped locations.

Cline: Continuous change in genetic composition of a population over a region (often an environmental gradient), such as a latitudinal or an altitudinal cline.

Coadapted: Combination of alleles at different genes that have higher fitness in combination than expected from the sum of their average effects, but harmful effects when recombined with alleles from another coadapted complex.

Coancestry: The coancestry of two individuals is the probability that two alleles, one from each individual, are identical by descent. Synonymous with kinship.

Common ancestor: An individual that is an ancestor of both parents of an individual.

Common garden experiment: Comparison of different genotypes contemporaneously in the same environment to distinguish genetic differences from environmental ones, usually for quantitative characters.

Conspecific: Belonging to the same species.

Corridor: Ribbon of habitat between population fragments.

cpDNA: See chloroplast DNA.

Critically endangered: A species with a very high probability of extinction within a short time, e.g. 50% probability of extinction within 10 years, or three generations, whichever is longer.

Decision tree: A diagram with a series of connected questions with dichotomous answers (yes or no) to aid in making decisions on some issue.

Demography: The study of how vital rates, such as fecundity, survival, and migration influence population growth and persistence.

Diploid: A genome with two doses of each chromosome (apart from sex chromosomes, if present).

Directional selection: Selection in which the most extreme high (or low) ranked individuals on some trait(s) from a population are most successful as parents of the next generation.

Diversifying selection: Selection of varying direction within the range of a species, e.g. favoring melanic peppered moths in polluted areas, but non-melanic forms in non-polluted areas.

Effective population size (N_e): The number of individuals that would result in the same loss of genetic diversity, inbreeding, or genetic drift if they behaved in the manner of an idealized population.

Endangered: A species or population with a high probability of extinction within a short time, e.g. a 20% probability of extinction within 20 years or 10 generations, whichever is longer.

Endemic: A population or species found in only one geographic area or country.

Equilibrium: The state where a population or other entity has no tendency to change from its present condition across time.

ESUs: See evolutionarily significant units.

Evolution: Change in the genetic composition of a population within or across generations.

Evolutionarily significant units (ESUs): Partially genetically differentiated populations that warrant status as separate units. There are diverse definitions that emphasize difference in neutral markers and adaptive differences to differing extents.

Evolutionary rescue: Demographic recovery of a population or species that is experiencing environmental change, through adaptive evolution, by utilizing pre-existing genetic diversity and/or new mutations.

Expected heterozygosity (H_e): The heterozygosity expected for a random mating population with given allele frequencies according to the Hardy–Weinberg equilibrium.

Extinction: Permanent demise of a population or species.

Extinction vortex: Describes the likely adverse interactions between human impacts, inbreeding, loss of genetic diversity, and demographic and environmental fluctuations that result in a reinforcing feedback loop and a downwards spiral in population size towards extinction.

F statistics: Measures of total inbreeding in a population (F_{IT}), partitioned into that due to inbreeding within fragments (F_{IS}) and that due to differentiation among them (F_{ST}).

Fitness: Reproductive fitness, the number of fertile offspring that survive to reproductive age, contributed by an individual (lifetime reproductive success).

Fixation: All individuals in a population being identically homozygous for a genetic marker or gene, e.g. all A_1A_1 in a diploid species.

Fixed gene differences: Populations homozygous for different alleles for one or more genetic markers or genes, or alternatively sharing no alleles at one or more markers.

Full-sib mating: A mating between two individuals having both parents in common, e.g. a mating between brother and sister.

Gene: A segment of the genome with a function.

Gene flow: Movement of alleles among populations via migrant individuals or gametes.

Gene trees (gene genealogies, phylogenies): Diagram showing the relationships among variants of a single genetic marker, typically derived from DNA sequences.

General Lineage Species Concept (GLSC): Concept that defines species as components of separately evolving metapopulation lineages.

Genetic connectivity: The degree to which gene flow alters evolutionary processes within and among population fragments.

Genetic diversity: The extent of genetic variation in a population or taxon, e.g. heterozygosity, or allelic diversity.

Genetic drift: Changes in the genetic composition of a population due to random sampling of alleles during reproduction in finite populations. Also referred to as random genetic drift.

Genetic erosion: The process whereby small populations lose genetic diversity and become inbred, leading to inbreeding depression, reduced ability to evolve, and elevated extinction risk.

Genetic load: The store of harmful alleles in a population or species.

Genetic marker: A segment of DNA, or simply inherited phenotypic variant that exhibits variation in a population or species.

Genetic rescue: Reduction in inbreeding, improvement in reproductive fitness, and increase in genetic diversity achieved by gene flow between distinct populations.

Genetic swamping: Substantial gene flow from another population, especially a related introduced species, resulting in a dilution of native genetic composition and potentially reduced local adaptation.

Genome: The entire DNA or all of the chromosomes in an individual, species, or organelle.

Habitat matrix: Habitat separating fragments of a population, as in cleared pasture land between forest fragments.

Haploid: Gamete or individual with one dose of each chromosome or genetic marker.

Hardy–Weinberg equilibrium: The equilibrium genotype frequencies achieved following random mating among individuals with no perturbation from mutation, gene flow, selection, or genetic drift. If two alleles A_1 and A_2 have frequencies of p and q, the Hardy–Weinberg equilibrium frequencies for the A_1A_1, A_1A_2, and A_2A_2 genotypes are p^2, $2pq$, and q^2, respectively, and are attained after one generation of random mating.

Hermaphrodite: An animal or plant with both sexes present in single individuals.

Heterosis: Hybrid vigor. Superior performance of hybrid genotypes for a quantitative character, usually superiority to both parents.

Heterozygosity: Possessing two different alleles at a genetic marker or gene in a diploid individual, or proportion of heterozygous individuals in a population.

Heterozygote advantage (overdominance): A form of selection where the heterozygote has a higher reproductive fitness than the homozygotes.

Idealized population: A conceptual random mating population with equal numbers of hermaphrodite individuals breeding in each generation, Poisson variation (mean = variance) in family sizes and no selection, gene flow, or mutation. Used as a standard against which other populations are compared when defining effective population sizes.

Identity by descent: Alleles that are identical copies of an allele present in a common ancestor.

Inbreeding: The production of offspring from mating of individuals related by descent, e.g. self-fertilization, brother–sister, cousin–cousin matings.

Inbreeding coefficient (F): The probability that two alleles for a genetic marker or gene in an individual are identical by descent. Used to measure the magnitude of inbreeding.

Inbreeding depression: Reduction in mean for a quantitative trait due to inbreeding, especially manifest by reproductive fitness traits.

***In situ* conservation:** Conservation of a species or population in its normal wild habitat.

Introgression: The incorporation of genes from one species (or sub-species) into the gene pool of another following hybridization.

Invasive species: A species introduced into a new range, that establishes and spreads, typically invading large areas and adversely affecting many native species.

Inversion: A chromosome aberration involving two breaks within a chromosome, turning the middle section through 180 degrees, and re-joining the ends such that gene order is changed, say from ABCD to ACBD.

Isolation by distance: Describes a distribution of population genetic composition where individuals show increasing genetic differentiation with increasing geographic distances between them.

Isolation by environment: Describes a distribution of population genetic composition where individuals show increasing genetic differentiation with increasing environmental differences between their locations.

IUCN: International Union for Conservation of Nature.

Karyotype: The number and appearance of the chromosomes in the nucleus of a cell.

Kinship coefficient (k_{ij}): The probability that two alleles, one from each of two individuals, are identical by descent (also termed coancestry). The inbreeding coefficient of hypothetical or actual progeny of the two individuals.

Kinship matrix: Table of pairwise kinship coefficients between different individuals or populations.

Lethal: Inconsistent with survival, as in a recessive lethal allele that results in death when homozygous.

Lethal equivalents (B): A measure for comparing the extent of inbreeding depression in different populations. One or more harmful alleles that would cause death or sterility if homozygous, e.g. one lethal or sterility allele, two alleles each with a 50% probability of causing death or sterility, etc. Typically estimated from the slope of the regression of the natural logarithm of survival, or relative fitness on the inbreeding coefficient F.

Lineage sorting: Random loss of genetic variants by genetic drift in different lineages deriving from a polymorphic common ancestral species (or population), resulting over generations in differentiated lineages.

Major histocompatibility complex (MHC): A large family of genes that play an important role in the vertebrate immune system.

Marker: see genetic marker.

Mating system: In plants, this refers to the proportion of mating events resulting from selfing compared to mating events between unrelated individuals (outcrossing). In animals, it refers to which males and females mate, degree of promiscuity, etc.

Mean kinship (mk): The average kinship of an individual with all individuals in a population, including itself. Minimizing mean kinship is the recommended method for genetically managing endangered species in captivity and we recommend its use in management of fragmented populations in the wild.

Meta-analysis: A statistical analysis based on the combined information from many different studies.

Metapopulation: A spatially distributed group of partially isolated population fragments of the same species that undergo local extinctions and recolonizations.

MHC: See major histocompatibility complex.

Microsatellite: A genetic marker with a short tandemly repeated DNA sequence (typically 1–5 bases in length), such as the sequence AC repeated 10 times. Typically shows variable number of repeats and high heterozygosities in populations.

Migration: Movement of individuals or gametes between populations. In conservation and evolutionary genetics, it is used to mean gene flow between populations.

Mitochondrial DNA (mtDNA): The circular DNA genome contained within mitochondria that codes for several proteins involved in energy metabolism and their expression machinery. Usually inherited from mothers, but not fathers, in animals and flowering plants.

Mixed mating: Populations or species that exhibit both selfing-fertilization and outcrossing, with selfing in the range of 20–80%.

Monomorphic: The presence of only one allele at a genetic marker or gene, generally taken to mean the most common allele is at a frequency of greater than 99% or 95%. Contrast with polymorphic.

Monophyletic: A group of species (or DNA sequences) that derive from the same common ancestral species (or DNA sequence). Converse is polyphyletic.

mtDNA: See mitochondrial DNA.

Mutation: A sudden genetic change, i.e. parents lack the allele, but it is present in one or more of their offspring, or a cell lacks the allele, but it is found in a daughter cell.

Mutation-selection balance: Equilibrium between the spontaneous occurrence of harmful mutations and natural selection removing them, resulting in low frequencies (typically < 1%) of harmful alleles of individual genes in populations. Occurs for many genes across the genome.

Natural selection: Differential mortality and/or reproduction among individuals in a population due to natural environmental processes that alters the genetic composition of a population if the differences are heritable.

Neutral mutation: A mutation that is equivalent in its effects on reproductive fitness to the existing allele.

Outbred: An individual whose parents are unrelated.

Outbreeding: A population or species that is not inbreeding, i.e. an approximately random mating population.

Outbreeding depression: Reduction in reproductive fitness due to crossing between two genetically divergent populations (or sub-species or species) caused by some combination of fixed chromosomal differences, disruption of local adaptation and gene complexes, and long isolation.

Outcrossing: Proportion of non-self matings in hermaphroditic species. Primarily outcrossing species have outcrossing rates of 80–100% (< 20% selfing). In animals, it may refer to crossing of an individual or population to another that is not closely related.

Parapatric: A geographic distribution of species or populations in which they abut.

Paraphyly: A group of individuals or species that includes an ancestor but not all of its descendants.

Pedigree: A chart specifying lines of descent and relationship among individuals.

Peripheral trait: A character with limited relationship to reproductive fitness, e.g. bristle number in fruit flies, tail length in rodents.

Phenotypic plasticity: Variation in the phenotype of individuals with the same genotype, as a response to different environments, e.g. change in hemoglobin levels following a move to high altitude for several months, or the production of different biomass by plants of the same genotype grown with different nutrient levels.

Phylogenetic Species Concepts (PSC): Defines species as diagnosably different population segments (variously based on fixed gene differences, lack of shared alleles, or reciprocal monophyly), irrespective of whether they are inherently reproductively isolated (compare Biological Species Concept).

Phylogenetic tree: A diagram representing the closeness of relationship between species or populations.

Polymorphic: The presence of more than one variant at a marker, gene, or chromosome, generally taken to mean the most common allele is at a frequency of less than 99% or 95% (compare with monomorphic).

Polyphyletic: A group of DNA sequences, individuals, or species that derives from more than one ancestral DNA sequence, individual, or species.

Polyploid: Having more than two doses of each chromosome, e.g. a tetraploid has four doses (4n).

Population: A group of individuals of the same species that could potentially interbreed with each other.

Population fragmentation: Conversion of a population's continuous distribution into separate spatial fragments through alteration of habitat, as in clearing for roads through a forest, or in partially clearing a forest for agriculture.

Population viability analysis (PVA): A systems modeling approach for predicting the fate of a population (including risk of extinction) due to the combined effects of systematic and variable threats faced by a population. Typically, population size, means and standard deviation of birth and death rates, density feedbacks, plus risks and severity of catastrophes, inbreeding depression, etc. are entered into a software program and many replicates projected over multiple generations using computer simulation. Used as a management and research tool in conservation biology.

Purging: Reducing the frequency of harmful alleles (frequently recessives ones) by natural selection, especially in populations that have suffered population size bottlenecks and/or inbreeding.

PVA: See population viability analysis.

Quantitative character: Typically, a trait with a continuous normal (bell-shaped) distribution influenced by genetic variation (typically at many genes) and environmental variation, e.g. fecundity, survival, height, behavior, and weight.

Random genetic drift: See genetic drift.

Random mating: A pattern of mating where the chances of two genotypes or phenotypes breeding is determined by their frequencies in the population (synonym: panmictic).

Reciprocal monophyly: Two groups of DNA sequences or populations (or taxa) with all members of the first more closely related to each other than to the second group, and vice versa.

Reintroduction: Returning a species or population to part of its former range where it had become extinct.

Relative fitness: The fitness of individuals or genotypes compared to that of others, e.g. if fitnesses of genotypes for a gene conferring warfarin resistance are 30%, 80%, and 54% for RR, RS, and SS genotypes, then their relative fitnesses are 30/80 = 0.375, 80/80 = 1, and 54/80 = 0.68, respectively.

Reproductive fitness: Lifetime reproductive success. The number of fertile offspring surviving to reproductive age, contributed by an individual. Encompasses mating ability, fertilization capacity, fecundity, and survival. Often referred to as fitness.

Selection coefficient (*s*): Difference in relative fitness between a genotype and the genotype with the highest fitness, e.g. if three genotypes A_1A_1, A_1A_2, and A_2A_2 have relative fitnesses of 1, 1, and 0.9, the selection coefficient for the A_2A_2 genotype is $s = 1 - 0.9 = 0.1$.

Selective sweep: Action of natural selection driving a single allele to fixation and at the same time reducing genetic diversity for surrounding DNA (typically not under selection itself). Common in genomic regions with low genetic recombination, such as mtDNA.

Self-incompatible: Genetically based inability of an individual (usually a plant) to produce offspring following attempted self-fertilization or mating with a close relative. Many plant species have genes that control self-incompatibility, and these typically have many alleles in large populations.

Selfing: Self-fertilizing.

Single nucleotide polymorphism (SNP): A position in the DNA of a species or population at which two or more alternative bases occur at appreciable frequency (> 1%).

SNP: See single nucleotide polymorphism.

Source–sink: A population structure in which permanent source populations supply individuals necessary to maintain otherwise-inviable sink populations.

Speciation: The processes by which populations diverge and become reproductively isolated, evolving into different species.

Species: Mayr defined species as "groups of actually or potentially interbreeding natural populations which are reproductively isolated from other such groups" according to the Biological Species Concept. There are many other definitions of species.

Statistical power: The ability to reject an erroneous null hypothesis.

Sub-species: Taxonomic units within species that are partially genetically differentiated populations, typically characterized by different phenotypes located in separate geographic regions.

Sympatric: The spatial arrangement of populations or species having the same or largely overlapping geographic distributions.

Tandem repeats: Multiple copies of the same DNA sequence lying one after another in a series, as in microsatellite repeats.

Taxa (singular taxon): One or more populations belonging to a taxonomic unit, e.g. several species or several sub-species.

Taxonomic Species Concept (TSC): A method of delineating species based on the determination of a taxonomist without specifying a defined species concept.

Tetraploid: Having four copies of each chromosome.

Threatened: A population or species that has a finite risk of extinction within a relatively short time frame, say greater than 10% risk of extinction within 100 years. Under the IUCN system this is the sum of the critically endangered, endangered, and vulnerable categories.

Translocation (physical movement): The movement of an individual animal or plant from one wild location to another as a result of human actions (distinguish from a chromosomal translocation).

Variance: The most commonly used measure of dispersion in a set of quantitative measurements. The average of the squared deviations from the mean. The square of the standard deviation.

Vulnerable: A species or population with a tangible risk of extinction within a moderate time, e.g. 10% probability within 100 years.

Wahlund effect: Reduction in heterozygosity, compared to Hardy–Weinberg expectations, in a population split into partially isolated population fragments (named after its discoverer).

References

These are the references new to this book, plus those required for copyright purposes. The references common to this book and *Genetic Management of Fragmented Animal and Plant Populations* are arranged in chapter order and available online at www.oup. co.uk/companion/FrankhamPG.

Aguilar, R., Quesada, M., Ashworth, L., et al., 2008. Genetic consequences of habitat fragmentation in plant populations: susceptible signals in plant traits and methodological approaches. Molecular Ecology 17, 5177–5188.

Allendorf, F.W., Luikart, G., Aitken, S.N., 2013. Conservation and the Genetics of Populations, 2nd edition. Wiley-Blackwell, Oxford, UK.

Anonymous, 2003. Wolves under threat as rabies outbreak bites. Nature 425, 892.

Anthonysamy, W.J.B., Dreslik, M.J., Douglas, M.R., et al., 2018. Population genetic evaluations within a co-distributed taxonomic group: a multi-species approach to conservation planning. Animal Conservation 21, 137–147.

Ballou, J.D., Lacy, R.C., 1995. Identifying genetically important individuals for management of genetic diversity in pedigreed populations. In: Population Management for Survival and Recovery: Analytical Methods and Strategies in Small Population Conservation, eds J. Ballou, M. Gilpin, T. Foose, pp. 76–111. Columbia University Press, New York.

Ballou, J.D., Lacy, R.C., Pollak, J.P., 2018. PMx: software for demographic and genetic analysis and management of pedigreed populations (version 1.5). Chicago Zoological Society, Brookfield, IL.

Banff National Park, 2012. Conservation Corridor. http://conservationcorridor. org/2012/10/banff-national-park/.

Barker, J.S.F., 2011. Effective population size of natural populations of *Drosophila buzzatii*, with a comparative evaluation of nine methods of estimation. Molecular Ecology 20, 4452–4471.

Barmentlo, S.H., Meirmans, P.G., Luijten, S.H., et al., 2018. Outbreeding depression and breeding system evolution in small, remnant populations of *Primula vulgaris*: consequences for genetic rescue. Conservation Genetics 19, 545–554.

Bay, R.A., Harrigan, R.J., Underwood, V.L., et al., 2018. Genomic signals of selection predict climate-driven population declines in a migratory bird. Science 359, 83–86.

Beheregaray, L.B., Sunnucks, P., Alpers, D.L., et al., 2000. A set of microsatellite loci for the hairy-nosed wombats (*Lasiorhinus krefftii* and *L. latifrons*). Conservation Genetics 1, 89–92.

Benson, J.F., Hostetler, J.A., Onorato, D.P., et al., 2011. Intentional genetic introgression influences survival of adults and subadults in a small, inbred felid population. Journal of Animal Ecology 80, 958–967.

Bohonak, A.J., 1999. Dispersal, gene flow and population structure. Quarterly Review of Biology 74, 21–45.

Booth, T.H., Nix, H.A., Busby, J.R., et al., 2014. BIOCLIM: the first species distribution modelling package, its early applications and relevance to most current MaxEnt studies. Diversity and Distributions 20, 1–9.

Bouzat, J.L., Cheng, H.H., Lewin, H.A., et al., 1998. Genetic evaluation of a demographic bottleneck in the greater prairie chicken. Conservation Biology 12, 836–843.

Buri, P., 1956. Gene frequency in small populations of mutant *Drosophila*. Evolution 10, 367–402.

Byers, O., Lees, C., Wilcken, J., et al., 2013. The One Plan Approach: the philosophy and implementation of CBSG's approach to integrated species conservation planning. WAZA Magazine 14, 2–5.

Ceballos, G., Ehrlich, P.R., Dirzo, R., 2017. Biological annihilation via the ongoing sixth mass extinction signaled by vertebrate population losses and declines. Proceedings of the National Academy of Sciences 114, E6089–E6096.

Charlesworth, D., Willis, J.H., 2009. The genetics of inbreeding depression. Nature Reviews Genetics 10, 783–796.

Chen, N., Cosgrove, E.J., Bowman, R., et al., 2016. Genomic consequences of population decline in the endangered Florida scrub-jay. Current Biology 26, 2974–2979.

Coates, D.J., Byrne, M., Moritz, C., 2018. Genetic diversity and conservation units: dealing with the species–population continuum in the age of genomics. Frontiers in Ecology and Evolution 6, 165, doi: 110.3389/fevo.2018.00165.

Coleman, R.A., Weeks, A.R., Hoffmann, A.A., 2013. Balancing genetic uniqueness and genetic variation in determining conservation and translocation strategies: a comprehensive case study of threatened dwarf galaxias, *Galaxiella pusilla* (Mack) (Pisces: Galaxiidae). Molecular Ecology 22, 1820–1835.

Coleman, R.A., Gauffre, B., Pavlova, A., et al., 2018. Artificial barriers prevent genetic recovery of small isolated populations of a low-mobility freshwater fish. Heredity 120, 515–532.

Cook, C., Sgrò, C., 2017. Aligning science and policy to achieve evolutionarily enlightened conservation management. Conservation Biology 31, 510–512.

Coulon, A., Fitzpatrick, J.W., Bowman, R., et al., 2008. Congruent population structure inferred from dispersal behaviour and intensive genetic surveys of the threatened Florida scrub-jay (*Aphelocoma cœrulescens*). Molecular Ecology 17, 1685–1701.

Coyer, J., Hoarau, G., Kjersti, S., et al., 2008. Being abundant is not enough: a decrease in effective population size over eight generations in a Norwegian population of the seaweed, *Fucus serratus*. Biology Letters 4, 755–757.

Coyne, J.A., Orr, H.A., 2004. Speciation. Sinauer, Sunderland, MA.

Cracraft, J., 1983. Species concepts and speciation analysis. In: Current Ornithology, ed. R.F. Johnston, pp. 159–187. Plenum Press, New York.

Cremer, K.W., 1966. Dissemination of seed from *Eucalyptus regnans*. Australian Forestry 30, 33–37.

Culver, M., Johnson, W.E., Pecon-Slattery, J., et al., 2000. Genomic ancestry of the American puma (*Puma concolor*). Journal of Heredity 91, 186–197.

Culver, M., Hedrick, P.W., Murphy, K., et al., 2008. Estimation of the bottleneck size in Florida panthers. Animal Conservation 11, 104–110.

De'ath, G., Fabricius, K.E., Sweatman, H., et al., 2012. The 27-year decline of coral cover on the Great Barrier Reef and its causes. Proceedings of the National Academy of Sciences 109, 17995–17999.

DeMay, S.M., Becker, P.A., Rachlow, J.L., et al., 2017. Genetic monitoring of an endangered species recovery: demographic and genetic trends for reintroduced pygmy rabbits (*Brachylagus idahoensis*). Journal of Mammalogy 98, 350–364.

de Queiroz, K., 2007. Species concepts and species delimitation. Systematic Biology 56, 879–886.

Dereeper, A., Guignon, V., Blanc, G., et al., 2008. Phylogeny.fr: robust phylogenetic analysis for the non-specialist. Nucleic Acids Research 36, W465–W469.

Dietz, J.M., Baker, A.J., Ballou, J.D., 2000. Demographic evidence of inbreeding depression in wild golden lion tamarins. In: Genetics, Demography and the Viability of Fragmented Populations, eds A.G. Young, G.M. Clarke, pp. 203–211. Cambridge University Press, Cambridge, UK.

Dolan, R.W., Yahr, R., Menges, E.S., et al., 1999. Conservation implications of genetic variation in three rare species endemic to Florida rosemary scrub. American Journal of Botany 86, 1556–1562.

Driscoll, C.A., Menotti-Raymond, M., Nelson, G., et al., 2002. Genomic microsatellites as evolutionary chronometers: a test in wild cats. Genome Research 12, 414–423.

Dudash, M.R., 1990. Relative fitness of selfed and outcrossed progeny in a self-compatible, protandrous species, *Sabatia angularis* L. (Gentianaceae): a comparison of three environments. Evolution 44, 1129–1139.

Du Plessis, S., Howard-McCombe, J., Melvin, Z., et al., 2018. Genetic diversity and cryptic population re-establishment: management implications for the Bojer's skink (*Gongylomorphus bojerii*). Conservation Genetics doi: org/10.1007/s10592-018-119-y.

Edmands, S., 2007. Between a rock and a hard place: evaluating the relative risks of inbreeding and outbreeding depression for conservation and management. Molecular Ecology 16, 463–475.

Eldridge, M.D.B., King, J.M., Loupis, A.K., et al., 1999. Unprecedented low levels of genetic variation and inbreeding depression in an island population of the black-footed rock-wallaby. Conservation Biology 13, 531–541.

El-Kassaby, Y.A., Yanchuk, A.D., 1994. Genetic diversity, differentiation, and inbreeding in Pacific yew from British Columbia. Journal of Heredity 85, 112–117.

Erickson, D.L., Fenster, C.B., 2006. Intraspecific hybridization and the recovery of fitness in the native legume *Chamaecrista fasciculata*. Evolution 60, 225–233.

Evenson, R.E., Gollin, D., 2003. Assessing the impact of the Green Revolution, 1960 to 2000. Science 300, 758–762.

Falconer, D.S., Mackay, T.F.C., 1996. Introduction to Quantitative Genetics, 4th edition. Longman, Harlow, England.

Falush, D., Stephens, M., Pritchard, J., K., 2007. Inference of population structure using multilocus genotype data: dominant markers and null alleles. Molecular Ecology Notes 7, 574–578.

Fenster, C.B., Ballou, J.D., Dudash, M.R., et al., 2018. Conservation and genetics. Yale Journal of Biology and Medicine 91, 491–501.

Finger, A., Kettle, C.J., Kaiser-Bunbury, C.N., et al., 2011. Back from the brink: potential for genetic rescue in a critically endangered tree. Molecular Ecology 20, 3773–3784.

Flesch, E.P., Rotella, J.J., Thomson, J.M., et al., 2018. Evaluating sample size to estimate genetic management metrics in the genomics era. Molecular Ecology Resources 18, 1077–1091.

Flescher, D., 2017. Florida panther shows sharp rebound. Sun Sentinel February 22, 2017.

Foden, W., Young, B., Akçakaya, H.R., et al., 2018. Climate change vulnerability assessment of species. WIREs Climate Change e551. ISSN 1757-7799.

Foose, T.J., 1986. Riders of the last ark: the role of captive breeding in conservation strategies. In: The Last Extinction, eds L. Kaufman, K. Mallory, pp. 141–165. MIT Press, Cambridge, MA.

Frankham, R., 1980. The founder effect and response to artificial selection in *Drosophila*. In: Selection Experiments in Laboratory and Domestic Animals, ed. A. Robertson, pp. 87–90. Commonwealth Agricultural Bureaux, Farnham Royal, UK.

Frankham, R., 1995. Effective population size/adult population size ratios in wildlife: a review. Genetical Research 66, 95–107.

Frankham, R., 2015. Genetic rescue of small inbred populations: meta-analysis reveals large and consistent benefits of gene flow. Molecular Ecology 24, 2610–2618.

Frankham, R., 2016. Genetic rescue benefits persist to at least the F3 generation, based on a meta-analysis. Biological Conservation 195, 33–36.

Frankham, R., 2018. Corrigendum to "Genetic rescue benefits persist to at least the F3 generation, based on a meta-analysis" [Biol. Conserv. 195 (2016) 33–36]. Biological Conservation 219, 174.

Frankham, R., Ballou, J.D., Briscoe, D.A., 2002. Introduction to Conservation Genetics. Cambridge University Press, Cambridge, UK.

Frankham, R., Ballou, J.D., Briscoe, D.A., 2004. A Primer of Conservation Genetics. Cambridge University Press, Cambridge, U.K.

Frankham, R., Ballou, J.D., Briscoe, D.A., 2010. Introduction to Conservation Genetics, 2nd edition. Cambridge University Press, Cambridge, U.K.

Frankham, R., Ballou, J.D., Eldridge, M.D.B., et al., 2011. Predicting the probability of outbreeding depression. Conservation Biology 25, 465–475.

Frankham, R., Ballou, J.D., Dudash, M.R., et al., 2012. Implications of different species concepts for conserving biodiversity. Biological Conservation 153, 25–31.

Frankham, R., Bradshaw, C.J.A., Brook, B.W., 2014. Genetics in conservation management: revised recommendations for the 50/500 rules, Red List criteria and population viability analyses. Biological Conservation 170, 56–63.

Frankham, R., Ballou, J.D., Ralls, K., et al., 2017. Genetic Management of Fragmented Animal and Plant Populations. Oxford University Press, Oxford, UK.

Fredrickson, R.J., Siminski, P., Woolf, M., et al., 2007. Genetic rescue and inbreeding depression in Mexican wolves. Proceedings of the Royal Society B: Biological Sciences 274, 2365–2371.

Friar, E.A., Ladoux, T., Roalson, E.H., et al., 2000. Microsatellite analysis of a population crash and bottleneck in the Mauna Kea silversword, *Argyroxiphium sandwicense* ssp.

sandwicense (Asteraceae), and its implications for reintroduction. Molecular Ecology 9, 2027–2034.

Frölicher, T.L., Fischer, E.M., Gruber, N., 2018. Marine heatwaves under global warming. Nature 560, 360–364.

Garnett, S.T., Christidis, L., 2017. Taxonomic anarchy hampers conservation. Nature 546, 25–27.

Gottelli, D., Sillero-Zubiri, C., Appelbaum, G.D., et al., 1994. Molecular genetics of the most endangered canid: the Ethiopian wolf *Canis simensis*. Molecular Ecology 3, 301–312.

Goudet, J., 2005. HIERFSTAT, a package for R to compute and test hierarchical F-statistics. Molecular Ecology Notes 5, 184–186.

Goudet, J., Kay, T., Weir, B.S., 2018. How to estimate kinship. Molecular Ecology 27, 4121–4135.

Grant, B.R., Grant, P.R., 2017. Watching speciation in action. Science 355, 910–911.

Guillot, G., Estoup, A., Mortier, F., et al., 2005. A spatial statistical model for landscape genetics. Genetics 170, 1261–1280.

Hall, B.G., 2017. Phylogenetic Trees Made Easy: A How-to Manual, 5th edition. Sinauer, Sunderland, MA.

Hamilton, J.A., Royauté, R., Wright, J.W., et al., 2017. Genetic conservation and management of the California endemic, Torrey pine (*Pinus torreyana* Parry): implications of genetic rescue in a genetically depauperate species. Ecology and Evolution 7, 7370–7381.

Hanski, I., Gilpin, M.E., 1997. Metapopulation Biology: Ecology, Genetics and Evolution. Academic Press, Orlando, FL.

Harrisson, K.A., Pavlova, A., Gonçalves da Silva, A., et al., 2016. Scope for genetic rescue of an endangered subspecies though re-establishing natural gene flow with another subspecies. Molecular Ecology 25, 1242–1258.

Harrisson, K.A., Amish, S.J., Pavlova, A., et al., 2017. Signatures of polygenic adaptation associated with climate across the range of a threatened fish species with high genetic connectivity. Molecular Ecology 26, 6253–6269.

Hasselgren, M., Angerbjörn, A., Eide, N.E., et al., 2018. Genetic rescue in an inbred Arctic fox (*Vulpes lagopus*) population. Proceedings of the Royal Society B: Biological Sciences 285, doi: 20172814.

Hauser, L., Carvalho, G.R., 2008. Paradigm shifts in marine fisheries genetics: ugly hypotheses slain by beautiful facts. Fish and Fisheries 9, 333–362.

Hedrick, P.W., 1983. Genetics of Populations. Science Books International, Boston, MA.

Hedrick, P.W., 1995. Gene flow and genetic restoration: the Florida panther as a case study. Conservation Biology 9, 996–1007.

Hedrick, P.W., Fredrickson, R., 2010. Genetic rescue guidelines with examples from Mexican wolves and Florida panthers. Conservation Genetics 11, 615–626.

Hedrick, P.W., Miller, P.S., Geffen, E., et al., 1997. Genetic evaluation of the three Mexican wolf lineages. Zoo Biology 16, 47–69.

Hereford, J., 2009. Postmating/prezygotic isolation, heterosis, and outbreeding depression in crosses within and between populations of *Diodia teres* (Rubiaceae) Walt. International Journal of Plant Sciences 170, 301–310.

Hoban, S., Gaggiotti, O., ConGRESS Consortium, et al., 2013. Sample Planning Optimization Tool for conservation and population Genetics (SPOTG): a software for choosing the appropriate number of markers and samples. Methods in Ecology and Evolution 4, 299–303.

Hoegh-Guldberg, O., Mumby, P.J., Hooten, A.J., et al., 2007. Coral reefs under rapid climate change and ocean acidification. Science 318, 1737–1742.

Hoffmann, A.A., Sgrò, C.M., 2011. Climate change and evolutionary adaptation. Nature 470, 479–485.

Hoffmann, A., Griffin, P., Dillon, S., et al., 2015. A framework for incorporating evolutionary genomics into biodiversity conservation and management. Climate Change Responses 2, doi: 10.1186/s40665-014-0009-x.

Hostetler, J.A., Onorato, D.P., Nichols, J.D., et al., 2010. Genetic introgression and the survival of Florida panther kittens. Biological Conservation 143, 2789–2796.

Hostetler, J.A., Onorato, D.P., Bolker, B.M., et al., 2012. Does genetic introgression improve female reproductive performance? A test on the endangered Florida panther. Oecologia 168, 289–300.

Houck, M.L., Lear, T.L., Charter, S.J., 2017. Animal cytogenetics. In: The AGT Cytogenetics Laboratory Manual, 4th edition, eds M.S. Arsham, M.J. Barch, H.J. Lawce, pp. 1055–1102. Wiley-Blackwell, Oxford, UK.

Hughes, A.L., 1991. MHC polymorphism and the design of captive breeding programs. Conservation Biology 5, 249–251.

Hughes, T.P., Kerry, J.T., Álvarez-Noriega, M., et al., 2017. Global warming and recurrent mass bleaching of corals. Nature 543, 373–377.

Hughes, T.P., Anderson, K.D., Connolly, S.R., et al., 2018. Spatial and temporal patterns of mass bleaching of corals in the Anthropocene. Science 359, 80–83.

Illinois Natural History Survey, 2016. Conservation guidance for Blanding's turtle (*Emydoidea blandingii*). Report prepared for the Illinois Department of Natural Resources, Division of Natural Heritage.

IPCC, 2014. Climate Change 2014: Synthesis Report. Contribution of Working Groups I, II and III to the Fifth Assessment Report of the Intergovernmental Panel on Climate Change [Core Writing Team, R.K. Pachauri and L.A. Meyer (eds).] IPCC, Geneva, Switzerland.

IUCN, 2018. IUCN Red List of Threatened Species. https://www.iucnredlist.org/.

IUCN/SSC, 2013. Guidelines for Reintroductions and Other Conservation Translocations. IUCN/SSC Reintroduction Specialist Group, Gland, Switzerland.

James, F., 1995. The status of the red-cockaded woodpecker in 1990 and the prospect for recovery. In: Red-Cockaded Woodpecker: Recovery, Ecology and Management, eds D.L. Kulhavy, R.G. Hooper, R. Costa, pp. 439–451. Center for Applied Studies, Stephen F. Austin State University, Nacogdoches, TX.

Johnson, R.N., O'Meally, D., Chen, Z., et al., 2018. Adaptation and conservation insights from the koala genome. Nature Genetics 50, 1102–1111.

Johnson, W.E., Onorato, D.P., Roelke, M.E., et al., 2010. Genetic restoration of the Florida panther. Science 329, 1641–1645.

Jump, A.S., Peñuelas, J., 2006. Genetic effects of chronic habitat fragmentation in a wind-pollinated tree. Proceedings of the National Academy of Sciences 103, 8096–8100.

Kahneman, D., 2011. Thinking, Fast and Slow. Farrar, Straus and Giroux, New York.

Kardos, M., Åkesson, M., Fountain, T., et al., 2018. Genomic consequences of intensive inbreeding in an isolated wolf population. Nature Ecology & Evolution 2, 124–131.

Kearns, A.M., Restani, M., Szabo, I., et al., 2018. Genomic evidence of speciation reversal in ravens. Nature Communications 9, 906.

Keller, M.C., Visscher, P.M., Goddard, M.E., 2011. Quantification of inbreeding due to distant ancestors and its detection using dense single nucleotide polymorphism data. Genetics 189, 237–249.

Kelly, E., Phillips, B.L., 2016. Targeted gene flow for conservation. Conservation Biology 30, 259–267.

Kelly, E., Phillips, B.L., 2019a. Targeted gene flow and rapid adaptation in an endangered marsupial. Conservation Biology 33, 112–121.

Kelly, E., Phillips, B.L., 2019b. How many, and when? Optimising targeted gene flow for a step change in the environment. Ecology Letters 22, 447–457.

King, R.B. 2013. Illinois conservation assessment for the Blanding's turtle (*Emydoidea blandingii*). Illinois Endangered Species Protection Board, Illinois Department of Natural Resources.

Kirov, I., Divashuk, M., Van Laere, K., et al., 2014. An easy "SteamDrop" method for high quality plant chromosome preparation. Molecular Cytogenetics 7, 1–10.

Knief, U., Hemmrich-Stanisak, G., Wittig, M., et al., 2015. Quantifying realized inbreeding in wild and captive animal populations. Heredity 114, 397–403.

Kronenberger, J.A., Funk, W.C., Smith, J.W., et al., 2017. Testing the demographic effects of divergent immigrants on small populations of Trinidadian guppies. Animal Conservation 20, 3–11.

Kulhavy, D.L., Hooper, R.G., Costa, R., 1995. Red-Cockaded Woodpecker: Recovery, Ecology and Management. Center for Applied Studies, Stephen F. Austin State University, Nacogdoches, TX.

Lacy, R.C., Pollak, J.P., 2014. Vortex: A Stochastic Simulation of the Extinction Process. Version 10.0. Chicago Zoological Society, Brookfield, IL.

Lacy, R.C., Ballou, J.D., Pollak, J.P., 2012. PMx: software package for demographic and genetic analysis and management of pedigreed populations. Methods in Ecology and Evolution 3, 433–437.

Lanzas, P., Perfectti, F., Garrido-Ramos, M.A., et al., 2018. Long-term monitoring of B-chromosome invasion and neutralization in a population of *Prospero autumnale* (Asparagaceae). Evolution 72, 1216–1224.

Larkin, M.A., Blackshields, G., Brown, N.P., et al., 2007. Clustal W and Clustal X version 2.0. Bioinformatics 23, 2947–2948.

Li, S., Li, B., Cheng, C., et al., 2014. Genomic signatures of near-extinction and rebirth of the crested ibis and other endangered bird species. Genome Biology 15, 557, doi: 10.1186/s13059-014-0557-1.

Madsen, T., Stille, B., Shine, R., 1996. Inbreeding depression in an isolated population of adders *Vipera berus*. Biological Conservation 75, 113–118.

Madsen, T., Ujvari, B., Olsson, M., 2004. Novel genes continue to enhance population growth in adders (*Vipera berus*). Biological Conservation 120, 145–147.

Martin, T.G., Burgman, M.A., Fidler, F., et al., 2012. Eliciting expert knowledge in conservation science. Conservation Biology 26, 29–38.

Matheny, K., 2018. 6–8 wolves to be relocated to help bolster Isle Royale pack. Detroit Free Press September 22, 2018, 6.

Mattila, A.L.K., Duplouy, A., Kirjokangas, M., et al., 2012. High genetic load in an old isolated butterfly population. Proceedings of the National Academy of Sciences 109, E2496–E2505.

Mayr, E., 1942. Systematics and the Origin of Species. Columbia University Press, New York.

Medina, I., Cooke, G.M., Ord, T.J., et al., 2018. Walk, swim or fly? Locomotor mode predicts genetic differentiation in vertebrates. Ecology Letters 21, 638–645.

Meffe, G.K., Carroll, C.R., 1997. Principles of Conservation Biology, 2nd edition. Sinauer, Sunderland, MA.

Mengel, R.M., Jackson, J.A., 1977. Geographic variation of the red-cockaded woodpecker. Condor 79, 349–355.

Morales, H.E., Pavlova, A., Amos, N., et al., 2018. Concordant divergence of mitogenomes and a mitonuclear gene cluster in bird lineages inhabiting different climates. Nature Ecology & Evolution 2, 1258–1267.

Murray, B.G., Young, A.G., 2001. Widespread chromosomal variation in the endangered grassland forb *Rutidosis leptorrhynchoides* F. Muell. (Asteraceae: Gnaphalieae). Annals of Botany 87, 83–90.

Mussmann, S.M., Douglas, M.R., Anthonysamy, W.J.B., et al., 2017. Genetic rescue, the greater prairie chicken and the problem of conservation reliance in the Anthropocene. Royal Society Open Science 4, 160736.

Nason, J.D., Ellstrand, N.C., 1995. Lifetime estimates of biparental inbreeding depression in the self-incompatible annual plant *Raphanus sativus*. Evolution 49, 307–316.

Nater, A., Mattle-Greminger, M.P., Nurcahyo, A., et al., 2017. Morphometric, behavioral, and genomic evidence for a new orangutan species. Current Biology 27, 3487–3498.

Nevo, E., Fu, Y.-B., Pavlicek, T., et al., 2012. Evolution of wild cereals during 28 years of global warming in Israel. Proceedings of the National Academy of Sciences 109, 3412–3415.

Newman, D., Tallmon, D.A., 2001. Experimental evidence for beneficial fitness effects of gene flow in recently isolated populations. Conservation Biology 15, 1054–1063.

Nguyen, L.-T., Schmidt, H.A., von Haeseler, A., et al., 2015. IQ-TREE: a fast and effective stochastic algorithm for estimating maximum-likelihood phylogenies. Molecular Biology and Evolution 32, 268–274.

Nieminen, M., Singer, M.C., Fortelius, W., et al., 2001. Experimental confirmation that inbreeding depression increases extinction risk in butterfly populations. American Naturalist 157, 237–244.

Nietlisbach, P., Keller, L.F., Camenisch, G., et al., 2017. Pedigree-based inbreeding coefficient explains more variation in fitness than heterozygosity at 160 microsatellites in a wild bird population. Proceedings of the Royal Society B: Biological Sciences 284, doi: 20162763.

Nietlisbach, P., Muff, S., Reid, J.M., et al., 2019. Nonequivalent lethal equivalents: models and inbreeding metrics for unbiased estimation of inbreeding load. Evolutionary Applications 12, 266–279.

NOAA, 2016. National Centers for Environmental Information. http://www.ncdc.noaa.gov/cag/.

Novak, B.J., Maloney, T., Phelan, R., 2018. Advancing a new toolkit for conservation: from science to policy. The CRISPR Journal 1, 11–15.

Oakley, C.G., Winn, A.A., 2012. Effects of population size and isolation on heterosis, mean fitness, and inbreeding depression in a perennial plant. New Phytologist 196, 261–270.

Oedekoven, K., 1980. The vanishing forest. Environmental Policy and Law 6, 184–185.

O'Grady, J.J., Brook, B.W., Reed, D.H., et al., 2006. Realistic levels of inbreeding depression strongly affect extinction risk in wild populations. Biological Conservation 133, 42–51.

Palstra, F.P., Fraser, D.J., 2012. Effective/census population size ratio estimation: a compendium and appraisal. Ecology and Evolution 2, 2357–2365.

Palstra, F.P., Ruzzante, D.E., 2008. Genetic estimates of contemporary effective population size: what can they tell us about the importance of genetic stochasticity for wild population persistence? Molecular Ecology 17, 3428–3447.

Pavlova, A., Beheregaray, L.B., Coleman, R., et al., 2017. Severe consequences of habitat fragmentation on genetic diversity of an endangered Australian freshwater fish: a call for assisted gene flow. Evolutionary Applications 10, 531–550.

Peakall, R., Smouse, P.E., 2006. GENALEX 6: genetic analysis in Excel. Population genetic software for teaching and research. Molecular Ecology Notes 6, 288–295.

Pelikan, S., Rogstad, S.H., 2013. NEWGARDEN: a computer program to model the population dynamics and genetics of establishing and fragmented plant populations. Conservation Genetics Resources 5, 857–862.

Pelletier, F., Coltman, D.W., 2018. Will human influences on evolutionary dynamics in the wild pervade the Anthropocene? BMC Biology 16, 7, doi: 10.1186/s12915-017-0476-1.

Pennisi, E., 2016. Shaking up the Tree of Life. Science 354, 817–821.

Pérez-Figueroa, A., Saura, M., Fernández, J., et al., 2009. METAPOP—a software for the management and analysis of subdivided populations in conservation programs. Conservation Genetics 10, 1097–1099.

Pfeifer, S.P., 2017. Direct estimate of the spontaneous germ line mutation rate in African green monkeys. Evolution 71, 2858–2870.

Phillips, S.J., Dudík, M., 2008. Modeling of species distributions with MAXENT: new extensions and a comprehensive evaluation. Ecography 31, 161–175.

Phillipsen, I.C., Kirk, E.H., Bogan, M.T., et al., 2015. Dispersal ability and habitat requirements determine landscape-level genetic patterns in desert aquatic insects. Molecular Ecology 24, 54–69.

Piel, W.H., Chan, L., Dominus, M.J., et al., 2009. TreeBASE v. 2: a database of phylogenetic knowledge. *e-BioSphere 2009*.

Pierson, J.C., Coates, D.J., Oostermeijer, J.G.B., et al., 2016. Consideration of genetic factors in threatened species recovery plans on three continents. Frontiers in Ecology and Environment 14, 433–440.

Popkin, G., 2018. Can a transgenic chestnut restore a forest icon? Science 361, 830–831.

Pray, L.A., Goodnight, C.J., Stevens, L., et al., 1996. The effect of population size on effective population size: an empirical study in the red flour beetle *Tribolium castaneum*. Genetical Research 68, 151–155.

Purcell, S., Neale, B., Todd-Brown, K., et al., 2007. PLINK: a tool set for whole-genome association and population-based linkage analyses. American Journal of Human Genetics 81, 559–575.

Ralls, K., Ballou, J., 1983. Extinction: lessons from zoos. In: Genetics and Conservation: A Reference for Managing Wild Animal and Plant Populations, eds C.M. Schonewald-Cox, S.M. Chambers, B. MacBryde, L. Thomas, pp. 164–184. Benjamin/Cummings, Menlo Park, CA.

Ralls, K., Ballou, J.D., Templeton, A., 1988. Estimates of lethal equivalents and the cost of inbreeding in mammals. Conservation Biology 2, 185–193.

Ralls, K., Ballou, J.D., Dudash, M.R., et al., 2018. Call for a paradigm shift in the genetic management of fragmented populations. Conservation Letters 11, e12412.

Razgour, O., Taggart, J.B., Manel, S., et al., 2018. An integrated framework to identify wildlife populations under threat from climate change. Molecular Ecology Resources 18, 18–31.

Reed, D.H., Lowe, E., Briscoe, D.A., et al., 2003. Inbreeding and extinction: effects of rate of inbreeding. Conservation Genetics 4, 405–410.

Rich, S.S., Bell, A.E., Wilson, S.P., 1979. Genetic drift in small populations of *Tribolium*. Evolution 33, 579–583.

Rieseberg, L.H., Willis, J.H., 2007. Plant speciation. Science 317, 910–914.

Robichaux, R.H., Friar, E.A., Mount, D.W., 1997. Molecular genetic consequences of a population bottleneck associated with reintroduction of the Mauna Kea silversword (*Argyroxiphium sandwicense* ssp. *sandwicense* [Asteraceae]). Conservation Biology 11, 1140–1146.

Roelke, M.E., Martenson, J., O'Brien, S.J., 1993. The consequences of demographic reduction and genetic depletion in the endangered Florida panther. Current Biology 3, 340–350.

Rousset, F., 2008. GENEPOP'007: a complete re-implementation of the GENEPOP software for Windows and Linux. Molecular Ecology Resources 8, 103–106.

Rowell, D.M., Lim, S.L., Grutzner, F., 2011. Chromosome analysis in invertebrates and vertebrates. In: Molecular Methods for Evolutionary Genetics, eds V. Orgogozo, M.V. Rockman, pp. 13–35. Humana Press, New York.

Rubin, C.S., Warner, R.E., Bouzat, J.L., et al., 2001. Population genetic structure of Blanding's turtles (*Emydoidea blandingii*) in an urban landscape. Biological Conservation 99, 323–330.

Rundle, H.D., Nagel, L., Boughman, J.W., et al., 2000. Natural selection and parallel speciation in sympatric stickleback. Science 287, 306–308.

Saccheri, I., Kuussaari, M., Kankare, M., et al., 1998. Inbreeding and extinction in a butterfly metapopulation. Nature 392, 491–494.

Scafaro, A.P., Gallé, A., Rie, J.V., et al., 2016. Heat tolerance in a wild *Oryza* species is attributed to maintenance of Rubisco activation by a thermally stable Rubisco activase ortholog. New Phytologist 211, 899–911.

Scheffers, B.R., De Meester, L., Bridge, T.C.L., et al., 2016. The broad footprint of climate change from genes to biomes to people. Science 354. aaf7671, doi: 10.1126/science.aaf7671.

Schlaepfer, D. R., Braschler, B., Rusterholz, H.-P., et al., 2018. Genetic effects of anthropogenic habitat fragmentation on remnant animal and plant populations: a meta-analysis. Ecosphere 9, e02488.

Schumaker, N.H., Brookes, A., 2018. HexSim: a modeling environment for ecology and conservation. Landscape Ecology 33, 197–211.

Sillero-Zubiri, C., Aungerbjorn, A., 2009. Arctic foxes and climate change. In: IUCN Red List. IUCN, Gland, Switzerland.

Somers, M.J., Hayward, M., eds., 2012. Fencing for Conservation: Restriction of Evolutionary Potential or a Riposte to Threatening Processes? Springer, New York.

Stockwell, D., 1999. The GARP modelling system: problems and solutions to automated spatial prediction. International Journal of Geographical Information Science 13, 143–158.

Sutherland, G.D., Harestad, A.S., Price, K., et al., 2000. Scaling of natal dispersal distances in terrestrial birds and mammals. Conservation Ecology 4, [online] http://www.consecol.org/vol4/iss1/art16/.

Szulkin, M., Garant, D., McCleery, R.H., et al., 2007. Inbreeding depression along a life-history continuum in the great tit. Journal of Evolutionary Biology 20, 1531–1543.

Taylor, A.C., Sherwin, W.B., Wayne, R.K., 1994. Genetic variation of microsatellite loci in a bottlenecked species: the northern hairy-nosed wombat *Lasiorhinus krefftii*. Molecular Ecology 3, 277–290.

Taylor, H., Dussex, N., van Heezik, Y., 2017. Bridging the conservation genetics gap by identifying barriers to implementation for conservation practitioners. Global Ecology and Conservation 10, 231–242.

Thompson, J.N., 2013. Relentless Evolution. University of Chicago Press, Chicago, IL.

Trinkel, M., Ferguson, N., Reid, A., et al., 2008. Translocating lions into an inbred lion population in the Hluhluwe-iMfolozi Park, South Africa. Animal Conservation 11, 138–143.

Ujvari, B., Klaassen, M., Raven, N., et al., 2018. Genetic diversity, inbreeding and cancer. Proceedings of the Royal Society B: Biological Sciences 285, 20172859.

Vale, P.F., Choisy, M., Froissart, R., et al., 2012. The distribution of mutational fitness effects of phage φx174 on different hosts. Evolution 66, 3495–3507.

van Oppen, M.J.H., Oliver, J.K., Putnam, H.M., et al., 2015. Building coral reef resilience through assisted evolution. Proceedings of the National Academy of Sciences 112, 2307–2313.

vonHoldt, B.M., Brzeski, K.E., Wilcove, D.S., et al., 2018. Redefining the role of admixture and genomics in species conservation. Conservation Letters 11, e12371.

Voyles, J., Woodhams, D.C., Saenz, V., et al., 2018. Shifts in disease dynamics in a tropical amphibian assemblage are not due to pathogen attenuation. Science 359, 1517–1519.

Waldvogel, A.-M., Wieser, A., Schell, T., et al., 2018. The genomic footprint of climate adaptation in *Chironomus riparius*. Molecular Ecology 27, 1439–1456.

Walling, C., Nussey, D., Morris, A., et al., 2011. Inbreeding depression in red deer calves. BMC Evolutionary Biology 11, 318.

Wang, J., 2011. COANCESTRY: a program for simulating, estimating and analysing relatedness and inbreeding coefficients. Molecular Ecology Resources 11, 141–145.

Waples, R.S., 1989. A generalized approach for estimating effective population size from temporal changes in allele frequency. Genetics 121, 379–391.

Waples, R.S., 2016. Making sense of genetic estimates of effective population size. Molecular Ecology 25, 4689–4691.

Waples, R.S., Gaggiotti, O., 2006. What is a population? An empirical evaluation of some genetic methods for identifying the number of gene pools and their degree of connectivity. Molecular Ecology 15, 1419–1439.

Waples, R.S., Luikart, G., Faulkner, J.R., et al., 2013. Simple life-history traits explain key effective population size ratios across diverse taxa. Proceedings of the Royal Society of London B: Biological Sciences 280, 20131339.

Weeks, A.R., Sgrò, C.M., Young, A.G., et al., 2011. Assessing the benefits and risks of translocations in changing environments: a genetic perspective. Evolutionary Applications 4, 709–725.

Weeks, A.R., Moro, D., Thavornkanlapachai, R., et al., 2015. Conserving and enhancing genetic diversity in translocation programs. In: Advances in Reintroduction Biology of Australian and New Zealand Fauna, eds D. Armstrong, M. Hayward, D. Moro, P. Seddon, pp. 127–140. CSIRO Publishing, Melbourne, Vic., Australia.

Weeks, A.R., Heinze, D., Perrin, L., et al., 2017. Genetic rescue increases fitness and aids rapid recovery of an endangered marsupial population. Nature Communications 8, 1071, doi: 10.1038/s41467-017-01182-3.

Weir, B.S., Goudet, J., 2017. A unified characterization of population structure and relatedness. Genetics 206, 2085–2103.

Westemeier, R.L., Brawn, J.D., Simpson, S.A., et al., 1998. Tracking the long-term decline and recovery of an isolated population. Science 282, 1695–1698.

White, L.C., Moseby, K.E., Thomson, V.A., et al., 2018. Long-term genetic consequences of mammal reintroductions into an Australian conservation reserve. Biological Conservation 219, 1–11.

Wilkins, J.S., 2009. Species: A History of the Idea. University of California Press, Berkeley, CA.

Wright, S., 1931. Evolution in Mendelian populations. Genetics 16, 97–159.

Wright, S., 1969. Evolution and the Genetics of Populations. 2. The Theory of Gene Frequencies. University of Chicago Press, Chicago, IL.

Wright, S., 1977. Evolution and the Genetics of Populations. 3. Experimental Results and Evolutionary Deductions. University of Chicago Press, Chicago, IL.

Zeller, M., Reusch, T.B.H., Lampert, W., 2008. Small effective population size in two planktonic freshwater copepod species (Eudiaptomus) with apparently large census sizes. Journal of Evolutionary Biology 21, 1755–1762.

Zhang, B., Li, M., Zhang, Z., et al., 2007. Genetic viability and population history of the giant panda, putting an end to the "evolutionary dead end"? Molecular Biology and Evolution 24, 1801–1810.

Index